Lecture N…
Mathematics

Edited by A. Dold and B. Eckmann

1392

Robert Silhol

Real Algebraic Surfaces

Springer-Verlag
Berlin Heidelberg New York London Paris Tokyo Hong Kong

Author

Robert Silhol
Institut de Mathématiques, Université des Sciences
et Techniques du Languedoc
34060 Montpellier Cedex, France

Mathematics Subject Classification (1980): 14 G 30, 14 J 26, 14 J 27, 14 J 28,
14 K 05, 14 K 10

ISBN 3-540-51563-1 Springer-Verlag Berlin Heidelberg New York
ISBN 0-387-51563-1 Springer-Verlag New York Berlin Heidelberg

© Springer-Verlag Berlin Heidelberg 1989
Printed in Germany

Printing and binding: Druckhaus Beltz, Hemsbach/Bergstr.
2146/3140-543210 – Printed on acid-free paper

INTRODUCTION

These notes are centred on one question : given a real algebraic surface X determine the topology of the real part $X(\mathbb{R})$.

Of course, since, to quote Hartshorne, the guiding problem in algebraic geometry is the classification problem (and this goes for real algebraic geometry also), the latter is very present in these notes. In fact it is present to the point, that we have only obtained a precise answer to our original question when we have obtained a precise answer to the classification problem. In this sense one could say that the underlying theme (and even, the main theme) of these notes is the classification problem of real algebraic surfaces.

This second preoccupation has dictated the plan of these notes and to some extent the methods used. We explain this. If two algebraic varieties are real isomorphic, then they certainly are complex isomorphic. Hence, our starting point, the well known Enriques-Kodaira classification of complex algebraic surfaces, and the plan.

To be able to make the most of the knowledge accumulated on complex algebraic surfaces we have used an alternative definition for real algebraic varieties, explicitly, we define them as complex algebraic varieties with an antiholomorphic involution. Otherwise said, we consider real algebraic varieties as complex algebraic varieties with an action of the Galois group $Gal(\mathbb{C}|\mathbb{R})$ (in the projective case, our only preoccupation, the two definitions are equivalent - see I.§1). This is the foundation of all the methods used in these notes.

From this point of view real algebraic surfaces fall into two classes, those for which the Galois action on $H^*(X(\mathbb{C}),\mathbb{Z})$ determines $H^*(X(\mathbb{R}),\mathbb{Z}/2)$ and those for which the Galois action only gives bounds on the dimensions of $H^*(X(\mathbb{R}),\mathbb{Z}/2)$. We call the first Galois-Maximal or GM-surfaces, they include rational surfaces, abelian surfaces, K3 surfaces and surfaces in $\mathbb{P}^3$. In the second class lie ruled surfaces and in general, surfaces fibred on a curve (elliptic surfaces... etc...).

We have been able to solve the classification problem completely for rational surfaces (chap. III and VI), abelian surfaces (chap. IV) and K3 surfaces (chap. VIII). These are all GM-surfaces.

For non-GM-surfaces we have concentrated on ruled and elliptic surfaces. The basic method used in both cases has been to study the Galois action on the fibration. We have obtained in this way complete results for ruled surfaces (chap. V) and complete local results for elliptic surfaces (chap. VII). We have also applied these methods to rational ruled surfaces (chap. VI).

The ideas, behind these methods are not new. The idea to consider a real algebraic variety as a complex variety with an antiholomorphic involution goes back at least to F. Klein and the idea to consider Galois action on cohomology is implicitly in Comessatti (although, of course, in different terms). The deep relations, of which we have made an essential use, between the cup-product form on $H^2(X(\mathbb{C}),\mathbb{Z})$ and $H^*(X(\mathbb{R}),\mathbb{Z}/2)$ are originally due to Arnold, Rokhlin, Gudkov, Kraknov and Kharlamov (see chap. II). To these authors is also due the use in real algebraic geometry of the Smith sequence. Finally, in this direction, the concept of GM-variety is due to Krasnov.

The idea to study Galois action on fibrations, goes back to Comessatti who studied the Galois action on pencils of curves, but the methods we have used are far closer to the methods developed by Manin and Iskovskih, the results of whom we have made an essential use.

We have tried to include in these notes most of the results on surfaces of the above mentioned authors.

Because of its importance it may be useful here, to say a few words on our treatment of the classification problem. As is usual in algebraic geometry such a problem divides into a discrete part (finding the right invariants) and a continuous part (the moduli problem).

For the discrete part, the first idea one can have, to consider complex invariants plus topological invariants of the real part, does not give a complete set of invariants. This is already true, and well known, in the case of curves. For GM-surfaces the correct set of invariants turns out to be the invariants for the action of the anti-holomorphic involution on $H^*(X(\mathbb{C}),\mathbb{Z})$ taken with the cup-product from and the Hodge decomposition (this includes all of the classical complex invariants). Note that for curves and abelian varieties, which are GM-varieties, this also gives the correct invariants.

We have taken these as our basic invariants even in the case of non-Galois-Maximal surfaces, in which case of course, they do not form a complete set of invariants.

For the continuous part, we have tried to build a Moduli space which classifies real algebraic surfaces with given invariants, up to real isomorphisms. This is somewhat different from considering the real part of the complex moduli space when it is known (for a discussion see chap. IV, §4). Even when a fine Moduli space does exist (for K3 surfaces for example) our method seems to give a more direct answer.

Among the results included in these notes, which are not specifically linked with the classification problem we should mention the results of chapter II giving bounds for the number of connected components and for the $h^1(X(\mathbb{R}),\mathbb{Z}/2)$ of a real algebraic surface and the results of chapter III on the subgroup of $H^1(X(\mathbb{R}),\mathbb{Z}/2)$ generated by algebraic cycles.

Among the subjects not treated in these notes the most important omission concerns surfaces of general type (they are only considered in a couple of examples). The main reason for this absence is that the study of complex surfaces of general type is still a field of active research and not all the questions we need an answer to (to apply our methods) have been solved. For specific families of surfaces of general type, for example double covers of rational surfaces, or surfaces fibred in curves of genus 2 (where we could have applied methods similar to those of chap. VII) the absence is essentially due to lack of time.

Another subject absent from these notes, is singular surfaces. The reasons for this omissions are more or less the same as the ones given for surfaces of general type.

A last subject omitted although originally planed to be included in these notes is the study of real automorphisms of surfaces and the parent problem of determining the number of distinct real structures on a complex surface.

Finally to end this introduction we would like to note that special attention has been given, throughout these notes, to examples. Indeed, we have tried to illustrate all of the important notions introduced in the text by adequate examples.

Prerequisites and notations

The general prerequisite for reading these notes, is a basic knowledge of algebraic geometry as exposed for example in Griffiths and Harris, Principles of Algebraic Geometry ([Gr & Ha]) chapters 0 and 1 or Hartshorne, Algebraic Geometry ([Ha]) chapters I and II, plus some knowledge on complex algebraic surfaces as exposed in [Gr & Ha] chapter 4 or [Ha] chapter V, or for more specific and precise results (but in this case we have tried to give complete references) Barth, Peters and Van de Ven [B & P & V], Beauville [Be], Shafarevich [Sha] or for K3-surfaces the Palaiseau Seminar [X].

On the other hand no knowledge of Real Algebraic Geometry is assumed. We expose the basic results needed in chapters I and II, and these notes can serve as an introduction to real algebraic geometry for non-specialists and graduate students in algebraic geometry.

The notations are the standard notations of algebraic geometry, for the less well known of these we have tried to give either definitions or precise references. There are some differences with the notations used by other authors. The most important are that, for reasons which will become clear in chapters I and II, we have used the Hodge numbers $h^{0,2}$ and $h^{0,1}$ in place of the geometric genus p_g and the irregularity q, $\chi(X(\mathbb{C}))$, topological Euler characteristic in place of $c_2(X)$ and $\chi(\mathcal{O}_X)$ for the Euler characteristic of the structure sheaf $\mathcal{O}_X$ (in place of $\chi(X)$ as used by some authors).

For notations concerning real algebraic geometry, we have systematically denoted G the Galois group $\mathrm{Gal}(\mathbb{C}|\mathbb{R})$ and S the generator of this group. This is why, although we consider in most cases S as an antiholomorphic involution on $X(\mathbb{C})$, we speak of the action of S on the groups $H^i(X(\mathbb{C}),\mathbb{Z})$, $H^i(X(\mathbb{C}),\mathbb{Q})$,... etc... (in place of S^*). For other important remarks on notations see I.(1.15).

TABLE OF CONTENTS

I. PRELIMINARIES ON REAL ALGEBRAIC VARIETIES.

1. Real structure on a complex algebraic variety.

We start by introducing the basic concepts we will be needing throughout these notes. We will introduce them in two different settings, complex analytic varieties on the one hand, schemes over $\mathbb{C}$ on the other. Of course we will not be using the full power of either of these theories but they provide a convenient frame for formulating some general definitions. The different definitions turn out to be equivalent when they concern the same objects, namely projective or quasi-projective algebraic varieties. We will take advantage of this and make constant use of the interplay between these two points of view.

Let f be a holomorphic function defined in a neighborhood of $z_0 \in \mathbb{C}^n$. Let $j_n : \mathbb{C}^n \longrightarrow \mathbb{C}^n$ and $j : \mathbb{C} \longrightarrow \mathbb{C}$ denote complex conjugation. We defined f^σ, <u>the conjugate of</u> f, to be the holomorphic function $j \circ f \circ j_n$ defined in a neighbourhood of $\bar{z}_0 = j_n(z_0)$.

In other words, if $z_0 = (z_{0,1},\dots,z_{0,n})$ and if f is defined by :

$$\sum_{k \in \mathbb{Z}^n} a_k(z_1-z_{0,1})^{k_1}\dots(z_n- z_{0,n})^{k_n}$$

in a neighbourhood of z_0, then f^σ if defined by :

$$f^\sigma(z) = \sum_{k \in \mathbb{Z}^n} \bar{a}_k(z_1-\bar{z}_{0,1})^{k_1}\dots(z_n-\bar{z}_{0,n})^{k_n}$$

(where we have used $\bar{\ }$'s to denote complexe conjugates) in a neighbourhood of $\bar{z}_0$.

Let $X(\mathbb{C})$ be a complex analytic variety in $\mathbb{C}^n$. We define the <u>com-</u>

<u>plex conjugate variety</u> $X^\sigma(\mathbb{C})$ to be $X^\sigma(\mathbb{C}) = \{z \,/\, j_n(z) = \bar{z} \in X(\mathbb{C})\}$. If U is an open set in $\mathbb{C}^n$ such that $X(\mathbb{C}) \cap U$ is the common zero locus of holomorphic functions $f_1,\ldots,f_m$ then $X^\sigma(\mathbb{C}) \cap \bar{U}$ (where $\bar{U} = \{z \,/\, \bar{z} \in U\}$ is the common zero locus of $f_1^\sigma,\ldots,f_m^\sigma$. We note that if the f_i's are polynomials then obviously the f_i^σ's are also polynomials.

In a more general way we can define this notion of conjugate variety globally for analytic varieties. Let $(X(\mathbb{C}),\mathcal{O}_{X(\mathbb{C})})$ ($\mathcal{O}_{X(\mathbb{C})}$ the sheaf of holomorphic functions) be a complexe analytic variety. We define <u>the complexe conjugate variety</u> to be $(X(\mathbb{C}),\bar{\mathcal{O}}_{X(\mathbb{C})})$, where $\bar{\mathcal{O}}_{X(\mathbb{C})}$ is the sheaf of antiholomorphic functions on $X(\mathbb{C})$.

To see that these two definitions are compatible assume $X(\mathbb{C}) \subset \mathbb{C}^n$. Then via j_n we can identify, as point sets, $X(\mathbb{C})$ and $X^\sigma(\mathbb{C})$. This identification identifies $\mathcal{O}_{X^\sigma(\mathbb{C})}$ with $\bar{\mathcal{O}}_{X(\mathbb{C})}$. The equivalence of the two definitions follows from this .

If $X(\mathbb{C})$ is a complex manifold we can reformulate the second definition in the following way : let (U_i,φ_i) be an atlas defining the complex structure on $X(\mathbb{C})$. We define $X^\sigma(\mathbb{C})$, <u>the complex conjugate manifold of</u> $X(\mathbb{C})$, to be defined by the atlas $(U_i,j_n\circ\varphi_i)$ where again $j_n : \mathbb{C}^n \longrightarrow \mathbb{C}^n$ is complex conjugation.

Of these three definitions the most useful, which we write separately, will be :

(1.1) <u>Définition</u> : <u>Let</u> $(X(\mathbb{C}),\mathcal{O}_{X(\mathbb{C})})$ <u>be a complex analytic variety</u>. <u>We will call</u> $(X(\mathbb{C}),\bar{\mathcal{O}}_{X(\mathbb{C})})$, <u>where</u> $\bar{\mathcal{O}}_{X(\mathbb{C})}$ <u>is the sheaf of anti-holomorphic function on</u> $X(\mathbb{C})$, <u>the complex conjugate variety of</u> $X(\mathbb{C})$. <u>When the sheaf</u> $\mathcal{O}_{X(\mathbb{C})}$ <u>is understood we will write simply</u> $X(\mathbb{C})$ <u>for the original complex analytic variety and</u> $X^\sigma(\mathbb{C})$ <u>for the complex conjugate variety</u>.

Our second point of view is the following : let X be a scheme over $\mathbb{C}$ and let $j : \mathbb{C} \longrightarrow \mathbb{C}$ be again complex conjugation. To X we can associate its <u>complex conjugate scheme</u> X^σ defined by composing the structural morphism $X \longrightarrow \mathrm{Spec}(\mathbb{C})$ with $j^* : \mathrm{Spec}(\mathbb{C}) \longrightarrow \mathrm{Spec}(\mathbb{C})$.

We note that if $\mathcal{O}_X$ (resp. $\mathcal{O}_{X^\sigma}$) is the sheaf of regular functions on X (resp. X^σ) then $\mathcal{O}_{X^\sigma} = \bar{\mathcal{O}}_X$ where $\bar{\mathcal{O}}_X(U) = \{j \circ f \,/\, f \in \mathcal{O}_X(U)\}$. This shows that this last definition is compatible with definition (1.1) .

In particular if X is projective (or quasi-projective) defined in some $\mathbb{P}^n(\mathbb{C})$ by polynomial equations $p_1(z) = \ldots = p_m(z) = 0$ then X^σ is defined by $p_1^\sigma(z) = \ldots = p_m^\sigma(z) = 0$ where p_i^σ is the conjugate polynomial of p_i.

If X is of finite type over $\mathbb{C}$, $X(\mathbb{C})$, its set of complex valued points, has a natural analytic structure. If $X^\sigma(\mathbb{C})$ is the set of complex points of X^σ, then, by the above, $X^\sigma(\mathbb{C})$ is the conjugate variety of $X(\mathbb{C})$ in the sense of definition (1.1).

(1.2) *Definition and Proposition* : *Let* X *be a scheme over* $\mathbb{C}$. *We will say that* (X,S) *or simply* S *is a real structure on* X, *if* S *is an involution on* X *such that the diagram*

$$\begin{array}{ccc} X & \xrightarrow{\ S\ } & X \\ \downarrow & & \downarrow \\ \mathrm{Spec}(\mathbb{C}) & \xrightarrow{\ j^*\ } & \mathrm{Spec}(\mathbb{C}) \end{array}$$

(where $j : \mathbb{C} \longrightarrow \mathbb{C}$ *is complex conjugation) commutes*.
If (X,S) *is a real structure on* X *and* $\mathcal{O}_X$ *is the structure sheaf then for any open set* U *of* X *the morphism* :

$$\begin{array}{ccc} \Gamma(U,\mathcal{O}_X) & \longrightarrow & \Gamma(S(U),\mathcal{O}_X) \\ f & \longmapsto & j \circ f \circ S = f^S \end{array}$$

is an isomorphism of rings.

Proof : By the definition, a real structure on X is a descent datum relative to the inclusion $\mathbb{R} \hookrightarrow \mathbb{C}$, in the sense of Grothendieck [Gr_2]. With this it is easy to check that, if S is a real structure, then $f \longmapsto f\circ S$ induces, for every open set U, an isomorphism $\mathcal{O}_X(U) \longrightarrow \mathcal{O}_{X^\sigma}(S(U))$, hence the last assertion of (1.2).

Let X be of finite type over $\mathbb{C}$ and consider $X(\mathbb{C})$, the set of complex points with its natural complex analytic structure. If (X,S) is a real structure, then S <u>restricted to</u> $X(\mathbb{C})$ <u>is just an anti-holomorphic involution on</u> $X(\mathbb{C})$. We have a partial converse to this :

(1.3) <u>Proposition</u> : <u>If</u> X <u>is a projective variety over</u> $\mathbb{C}$ <u>then</u> X <u>has a real structure if and only if there exists an anti-holomorphic involution on</u> $X(\mathbb{C})$, <u>the set of complex points of</u> X.

<u>Proof</u> : We only need to prove the "if" part. If X is projective then clearly the conjugate variety is also projective. Let S be an anti-holomorphic involution on $X(\mathbb{C})$ and $\sigma : X(\mathbb{C}) \longrightarrow X^\sigma(\mathbb{C})$ the canonical map induced by the identity on the point sets. The map $S\circ\sigma$ is holomorphic, hence, since X is projective, algebraic by GAGA [Se1]. In other words, S induces a continuous involution on X. Identifying X and X^σ, as point sets we can write that S induces, for all open sets U of X, an isomorphism $\mathcal{O}_X(U) \longrightarrow \mathcal{O}_{X^\sigma}(S(U))$. Since by definition the map $f \mapsto j\circ f$ identifies $\mathcal{O}_{X^\sigma}(S(U))$ and $\mathcal{O}_X(S(U))$, we see that S satisfies the conditions of (1.2).

Let X be a scheme over $\mathbb{C}$. We will say that X has a real model if there exists a scheme X_0 over $\mathbb{R}$ such that $X \cong X_0 \times_{\mathbb{R}} \mathbb{C}$. We will say, in such a case, that X_0 is a <u>real model</u> of X.

If a scheme over $\mathbb{C}$ has a real model then the action of the Galois group $Gal(\mathbb{C}|\mathbb{R})$ on X, defines in a natural way a real structure on X. We have a converse to this :

(1.4) <u>Proposition</u> : <u>Let</u> X <u>be a projective or quasi-projective scheme over</u> $\mathbb{C}$. X <u>has a real model if and only if</u> X <u>admits a real structure</u> (X,S). <u>More precisely there exists a real structure</u> (X,S) <u>on</u> X <u>if and only if there exists a real model</u> X_0 <u>for</u> X <u>and an isomorphism</u>

$\varphi : X \longrightarrow X_o \times_{\mathbb{R}} \mathbb{C}$ such that $S = \varphi^{-1} \circ \sigma \circ \varphi$, where σ is induced by complex conjugation in $X_o \times_{\mathbb{R}} \mathbb{C}$. For a fixed (X,S), φ and X_o are unique up to real isomorphism.

Proof : This is nothing but a reformulation, in our special case, of a well known theorem of Weil (see [We_1] or [Gr_2] Exp. 190 Théorème 3).

The trivial example given by $\mathbb{P}^1_{\mathbb{R}}$ and the curve defined in $\mathbb{P}^2$ by $x^2+y^2+z^2 = 0$ shows that complex varieties can have different non real-isomorphic real models. (1.4) implies in such a case that they have different real structures. If we have two real structures (X,S) and (X,S'), (1.4) implies that they correspond to a same real model X_o if and only if there exists a complex automorphism φ of X such that :

(1.5) $$S = \varphi^{-1} \circ S' \circ \varphi .$$

As a consequence, we will say that two real structures (X,S) and (X',S') are isomorphic or real equivalent if there exists an isomorphism $\varphi : X \longrightarrow X'$ such that S, S' and φ verify (1.5).

Let A be a set and G a group operating on A. We will denote A^G the set of fixed points of A under the action of G. Let X_o be a real algebraic variety (or more generally a scheme over $\mathbb{R}$). Let $G = \mathrm{Gal}(\mathbb{C}|\mathbb{R})$. We have :

(1.6) $$X_o(\mathbb{R}) = X_o(\mathbb{C})^G .$$

Let (X,S) be a real structure on a complex algebraic variety. We will call $X(\mathbb{C})^G$ (where, as always, $G = \mathrm{Gal}(\mathbb{C}|\mathbb{R})$ acts on $X(\mathbb{C})$ through S) the real part of X. We will write $X(\mathbb{R})$ or $(X,S)(\mathbb{R})$, if the emphasis is on S, for this real part.

We will assume until the end of this § that X is a complex algebraic variety and that X has a real structure (X,S).

(1.7) Let $\mathcal{L}$ be a quasi-coherent sheaf of $\mathcal{O}_X$-modules and U an affine open set in X. Let $\mathcal{L}(S(U)) = M$, M an $\mathcal{O}_X(S(U))$-module. Define M^S to be the $\mathcal{O}_X(U)$-module with same underlying additive group as M but exterior multiplication defined by $(f,m) \mapsto (j\circ f\circ S)m$ (product taken in M). <u>We define the sheaf of</u> $\mathcal{O}_X$<u>-modules</u> $\mathcal{L}^S$ <u>by setting</u> $\mathcal{L}^S(U) = M^S$.

(1.7.1) If $\mathcal{L}$ is a sheaf of functions with, say, values in $\mathbb{C}^n$, we can describe $\mathcal{L}^S$ as defined by $\mathcal{L}^S(U) = \{j\circ f\circ S \ / \ f \in \mathcal{L}(S(U))\}$.

We will say that a quasi-coherent sheaf of $\mathcal{O}_X$-modules, $\mathcal{L}$, <u>is real</u> or S-<u>real</u> if : $\mathcal{L} = \mathcal{L}^S$ (equality not isomorphism).

This definition is motivated by the following result :

(1.8). <u>Lemma</u> : <u>Let</u> $X = X_0 \times_{\mathbb{R}} \mathbb{C} \xrightarrow{\ p\ } X_0$ (p <u>projection on the first factor</u>). <u>Then</u> $\mathcal{L}$ <u>is real</u>, <u>for the canonical real structure on</u> X, <u>if and only if there exists</u> $\mathcal{L}_0$ <u>such that</u> $\mathcal{L} = p^*\mathcal{L}_0$ (<u>or to be more precise</u> $p^{-1}(\mathcal{L}_0) \otimes_{\mathcal{O}_{X_0}} \mathcal{O}_X = \mathcal{L}$).

<u>Proof</u> : (This is a special case of a far more general result of Grothendieck). The question is local, so we can reduce to the affine case. Let $X_0 = \mathrm{Spec}(A)$, $X = \mathrm{Spec}(A \otimes_{\mathbb{R}} \mathbb{C})$ and $\mathcal{L} = \tilde{M}$ ($\tilde{M}$ the sheaf canonically associated to M). By hypothesis S induces an automorphism on M, hence a Galois action. Considering the A-module M^G we have : $M = M^G \otimes_A (A \otimes_{\mathbb{R}} \mathbb{C})$. Writing $\mathcal{L}_0 = \tilde{M}^G$ we get the result.

If we consider $\mathcal{L}$ as a sheaf of $\mathbb{C}$-vector spaces, the canonical map $\mathcal{L}(U) \longrightarrow \mathcal{L}^S(S(U))$ is an anti-isomorphism of $\mathbb{C}$-vector spaces. The same is also true for $\mathcal{L}_x$ and $\mathcal{L}^S_{S(x)}$. We therefore have :

$$\dim_{\mathbb{C}} \mathcal{L}(U) = \dim_{\mathbb{C}}(\mathcal{L}^S(S(U)) \text{ and } \dim_{\mathbb{C}}\mathcal{L}_x = \dim_{\mathbb{C}}\mathcal{L}^S_{S(x)} \ .$$

It is easy to see that the functor $\mathcal{L} \longrightarrow \mathcal{L}^S$ is exact and transforms injective objects into injective objects. We can thus generalize the above equalities into :

(1.9) $$\dim_{\mathbb{C}} H^i(X,\mathcal{L}) = \dim_{\mathbb{C}} H^i(X,\mathcal{L}^S) \quad .$$

(1.9.1) Let Y be a sub-scheme of X defined by a sheaf of ideals $\mathfrak{J}$. S(Y) is naturally a sub-scheme of X and by (1.7.1) this scheme is defined by the sheaf $\mathfrak{J}^S$.

Let Y be an algebraic variety over $\mathbb{C}$ (that is, an integral separated scheme of finite type over $\mathbb{C}$). Then we can, in the usual way define

$$p_a(Y) = (-1)^{\dim Y}(\chi(Y,\mathcal{O}_Y)) \qquad \text{(arithmetic genus)}$$

$$q(Y) = \dim_{\mathbb{C}} H^1(Y,\mathcal{O}_Y). \qquad \text{(irregularity)}$$

By (1.9) and the preceding remark we have, writing Y^S for S(Y) :

(1.10) $$p_a(Y) = p_a(Y^S)$$

$$q(Y) = q(Y^S) \quad .$$

We end this § with a few remarks on local parameter systems. We continue to assume that X has a real structure (X,S). Let us assume moreover that X is smooth and of dimension n.

Using the isomorphism of (1.2) we can always choose local systems of parameters $(\varphi_1,\dots,\varphi_n)$ on X such that :

(1.11) $$\forall\, x \in X \qquad (\varphi_{i,x})^S = \varphi_{i,S(x)} \quad ,$$

where $(\varphi_{i,x})^S$ is as defined in (1.2). We will in particular always choose local systems in this way when we consider the complex analytic structure of $X(\mathbb{C})$ and $(\varphi_1,\dots,\varphi_n)$ as analytic parameters.

Considering the underlying C^∞ structure of $X(\mathbb{C})$, there is one important point to note in this context. Let $\{\varphi_i\}$ be an analytic system satisfying (1.11). Then the action of S alone on this system is given by :

(1.12) $$\varphi_{i,x} \circ S = j \circ \varphi_{i,S(x)} = \bar{\varphi}_{i,S(x)} \quad .$$

This means in particular that if $(z_1,\ldots,z_n)$ is a local system satisfying (1.11) and we consider $(z_1,\bar{z}_1,\ldots,z_n,\bar{z}_n)$ as local $\mathbb{C}$-valued coordinates for the C^∞ structure, then the action of S on this last system transforms it into

(1.13) $$(\bar{z}_1, z_1, \ldots, \bar{z}_n, z_n) \quad .$$

A last remark that imposes itself here, is that if we have made convention (1.11) and if x is a smooth real point of $X(\mathbb{C})$, then the restriction to $X(\mathbb{R})$ of the $\varphi_{i,x}$ form a system of real analytic parameters of $X(\mathbb{R})$ in the neighbourhood of x. In particular we have as a consequence the classical result :

(1.14) <u>If</u> $X(\mathbb{C})$ <u>is smooth and of complex dimension</u> n <u>and if</u> $X(\mathbb{R})$ <u>is non empty, then</u> $X(\mathbb{R})$ <u>is smooth of real dimension</u> n.

(1.15) <u>Remarks on notations</u> : We will in the sequel make free use of the results of §1. We specify here in what way.

We will always assume from now on that X is a projective or quasi-projective algebraic variety over $\mathbb{C}$.

If (X,S) is a real structure on X we will say that (X,S) or X, if S is understood, is a <u>real algebraic variety</u>. This is a mild abuse of notations consisting in identifying, by way of (1.4), X with its real model. In fact since we are primarily interested in $X(\mathbb{C})$ and $X(\mathbb{R})$, both well defined by X and (X,S), this abuse will be in practice of no consequence (in the sense that if $X \cong X_0 \times_{\mathbb{R}} \mathbb{C}$, then $X(\mathbb{R}) \cong X_0(\mathbb{R})$ ((1.6)) and $X(\mathbb{C}) \cong X_0(\mathbb{C})$).

If X is projective and S is <u>an anti-holomorphic involution on</u> $X(\mathbb{C})$, we will use (1.3) and say that (X,S) <u>is a real structure on</u> X.

We will also need, for some results, to consider "real varieties" as schemes over $\mathbb{R}$. In this case we will speak of varieties over $\mathbb{R}$ (curve over $\mathbb{R}$, surface over $\mathbb{R}$, ...) and use the notation $X_0, Y_0, \ldots$. In this case S will denote the involution induced by Galois action on

$X = X_o \times_{\mathbb{R}} \mathbb{C}$. We hope that no confusions will result from this use of two languages.

2. Galois action on cohomology and on the Hodge decomposition.

In what follows we assume that X is a smooth, irreducible, projective algebraic variety of the form $X_o \times_{\mathbb{R}} \mathbb{C}$, where X_o is an algebraic variety over $\mathbb{R}$ in the scheme sense. We will consider on X the canonical real structure (X,S) induced by the Galois action on $X_o \times_{\mathbb{R}} \mathbb{C}$ and say (see (1.15)) that X is a real variety.

Note that the hypotheses on X imply that X_o is also irreducible, smooth and projective. Irreducible and smooth by general descent theory and projective because if $\mathcal{L}$ is a very ample sheaf on X, then $\mathcal{L} \otimes \mathcal{L}^S$ is a real very ample sheaf.

On the other hand, if we had started with X_o irreducible smooth and projective, X would be smooth and projective but not necessarily irreducible (X may not be connected - consider $\mathbb{R}[X]/X^2+1$). Our hypothesis on X is equivalent to X_o geometrically irreducible.

Let X be, as above, a real variety. The Galois group $\mathrm{Gal}(\mathbb{C}|\mathbb{R}) = G = \{1,S\}$ operates continuously and properly on X. It also opertates in the same way on $X(\mathbb{C})$ and hence on the groups $H^i(X(\mathbb{C}),\mathbb{Z})$, $H^i(X(\mathbb{C}),\mathbb{Q})$,... etc... . The aim of the following two § is to establish links between this action and the groups $H^i(X(\mathbb{R}),\mathbb{Z}/2)$.

Our first result in this direction is :

(2.1) Proposition : *If X is a smooth projective real variety, then we have for the Euler characteristic of* $X(\mathbb{R})$:

$$\chi(X(\mathbb{R})) = \sum_{\substack{i \equiv 0 \\ (\mathrm{mod}.2)}} (2b_i - B_i)$$

where B_i *is the i-th Betti number of* $X(\mathbb{C})$, $b_i = \dim((H^i(X(\mathbb{C}),\mathbb{Q})^G)$ *and* $\chi(X(\mathbb{R})) = \Sigma\, (-1)^i \dim H^i(X(\mathbb{R}),\mathbb{Q})$ *is the topological Euler characteristic.*

Proof : Recall that $X(\mathbb{R}) = X(\mathbb{C})^G$, the fixed part of $X(\mathbb{C})$ under the action of G. We will write $Y = X(\mathbb{C})/G$ for the topological quotient and $\pi : X(\mathbb{C}) \longrightarrow Y$ for the canonical surjection.

We can always find a finite triangulation of $X(\mathbb{C})$ such that $X(\mathbb{R})$ forms a sub-complex of this triangulation. Using this construction it is easy to prove (see Floyd [Fl_1] p. 138 for a proof in a more general context) that :

$$(2.2) \qquad \chi(X(\mathbb{C})) = 2\chi(Y) - \chi(X(\mathbb{R})),$$

where $\chi(X(\mathbb{C}))$, $\chi(Y)$ are the topological Euler characteristics of $X(\mathbb{C})$ and Y.

Considering the morphism π^* induced by the surjection π we have an isomorphism :

$$(2.3) \qquad H^i(Y,\mathbb{Q}) \cong (H^i(X(\mathbb{C}),\mathbb{Q}))^G$$

(see [Gr_1] p. 202 or [Fl_2] p. 38).

Recalling that $\chi(X(\mathbb{C})) = \Sigma\, (-1)^i \dim H^i(X(\mathbb{C}),\mathbb{Q})$ and $\chi(Y) = \Sigma(-1)^i \dim H^i(Y,\mathbb{Q})$, we can combine (2.2) and (2.3) to obtain, with the notations of (2.1) :

$$\chi(X(\mathbb{R})) = \sum_{i=0}^{2n} (2b_i - B_i).$$

To prove (2.1) we only need to prove that for i odd : $2b_i = B_i$. This will follow from the more general result :

(2.4) Lemma : Let $F_\infty = S^* \otimes \mathrm{Id}$ operate on $H^*(X(\mathbb{C}),\mathbb{Q}) \otimes_{\mathbb{Q}} \mathbb{C} = H^*(X(\mathbb{C}),\mathbb{C})$. Then $F_\infty H^{p,q}(X(\mathbb{C})) = H^{q,p}(X(\mathbb{C}))$, where the $H^{p,q}(X(\mathbb{C}))$ are the sub-spaces obtained in the Hodge decomposition of $H^*(X(\mathbb{C}),\mathbb{C})$.

Proof : Let $\omega = \Sigma\, \omega_{IJ}\, dz_I \wedge d\bar{z}_J$ (I and J multi-indices and

$dz_I = dz_{i_1} \wedge \ldots \wedge dz_{i_n}$) a C^∞ differential form on $X(\mathbb{C})$. Using convention (1.11) for the local coordinates and recalling (1.13), we get :

$$F_\infty \omega = \Sigma\, \omega_{IJ} \circ S\, d\bar{z}_I \wedge dz_J .$$

In other words, F_∞ transforms forms of type (p,q) into forms of type (q,p).

For a general C^∞ function on $\mathbb{C}^n$ we have :

$$\frac{\partial}{\partial \bar{z}_i} (\varphi \circ j_n) = \left(\frac{\partial}{\partial z_i} \varphi \right) \circ j_n .$$

In $X(\mathbb{C})$ we get the same formula replacing j_n by S.

Recalling that $\partial \omega = \Sigma \frac{\partial}{\partial z_i} \omega_{IJ}\, dz_i \wedge dz_I \wedge d\bar{z}_J$ and $\bar{\partial}\omega = \Sigma \frac{\partial}{\partial \bar{z}_j} \omega_{IJ}\, d\bar{z}_i \wedge dz_I \wedge d\bar{z}_J$ we get, combining with the above result

$$F_\infty \partial = \bar{\partial} F_\infty \text{ and } F_\infty \bar{\partial} = \partial F_\infty .$$

Writing $d = \partial + \bar{\partial}$, we see that F_∞ transform d-closed (resp. d-exact) forms into d-closed (resp. d-exact) forms.

Writing $Z_d^{p,q}(X(\mathbb{C}))$ for the space of d-closed forms of type (p,q) and $A^*(X(\mathbb{C}))$ for the space of $\mathbb{C}$-valued differential forms, we have :

$$H^{p,q}(X(\mathbb{C})) = Z_d^{p,q}(X(\mathbb{C}))/dA^*(X(\mathbb{C})) \cap Z_d^{p,q}(X(\mathbb{C})) .$$

The above remarks prove that $F_\infty H^{p,q}(X(\mathbb{C})) = H^{q,p}(X(\mathbb{C}))$ and hence (2.4).

Another way of proving (2.4), that is interesting in its own right, is to note that F_∞ transforms harmonic forms into harmonic forms. To see this note that by the above, we have $F_\infty d = dF_\infty$ and that if $*\omega = c_{p,q}\, \Sigma\, \bar{\omega}_{IJ} dz_{I'} \wedge d\bar{z}_{J'}$ ($c_{p,q}$ a real constant depending only on (p,q) - for more details see [Gr & Ha] p. 82 - and $I' = \{1,\ldots,n\} \backslash I$), then $F_\infty * = *F_\infty$. Writing $d^* = -*d*$ and $\Delta_d = dd^*+d^*d$ we get $F_\infty d^* = d^*F_\infty$ and $\Delta_d F_\infty = F_\infty\Delta_d$. Hence the assertion by [Gr & Ha] p. 115, since by hypothesis we are on a compact Kähler Manifold.

(2.4) implies in particular that an element of $\bigoplus_{\substack{p+q=i\\ p\neq q}} H^{p,q}$ is fixed by F_∞ if and only if it is of the form $\omega + F_\infty\omega$. In other words, the dimension of the sub-space of $\bigoplus_{\substack{p+q=i\\ p\neq q}} H^{p,q}$ fixed by F_∞ is equal to

$\sum_{\substack{p+q=i\\ p<q}} h^{p,q}$ where $h^{p,q}$ is the dimension of $H^{p,q}$.

<u>If</u> i <u>is odd</u> this implies :

$$b_i = (1/2)B_i \quad . \tag{2.5}$$

This ends the proof of (2.1).

<u>If</u> i <u>is even</u> the lemma implies :

$$b_i \geqslant \sum_{\substack{p+q=i\\ p<q}} h^{p,q} \quad . \tag{2.6}$$

Because $H^i(X(\mathbb{C}),\mathbb{C})$ is a vector space and F_∞ is an involution, it decomposes into the direct sum of the fixed part under F_∞ and the fixed part under $-F_\infty$. The lemma thus also implies :

$$b_i \leqslant B_i - \sum_{\substack{p+q=i\\ p<q}} h^{p,q} \quad . \tag{2.7}$$

<u>Remark</u> : Clearly all the results of this § continue to hold if $X(\mathbb{C})$ is a compact Kähler manifold and S an anti-holomorphic involution on $X(\mathbb{C})$.

3. Galois action on cohomology, the Smith sequence and the Harnack-Thom-Krasnov inequalities.

We will use the notations of §2. In particular, let again $Y = X(\mathbb{C})/G$ and $\pi : X(\mathbb{C}) \longrightarrow Y$ be the canonical surjection. We consider on Y the sheaf $\mathcal{F} = \pi_*(\mathbb{Z}/2)$ (where we write $\mathbb{Z}/2$ for the constant sheaf with stalk $\mathbb{Z}/2$).

$G = \{1,S\}$ operates on $\mathcal{F}$ and we can consider the sheaf $(1+S)\mathcal{F}$. We are going to prove that the following sequence is an exact sequence of sheaves on Y :

$$(3.0) \qquad 0 \longrightarrow (1+S)\mathcal{F} \longrightarrow \mathcal{F} \xrightarrow{(1+S)\oplus r} (1+S)\mathcal{F}\oplus\mathcal{F}_{|X(\mathbb{R})} \longrightarrow 0$$

where r is the restriction morphism and $\mathcal{F}_{|X(\mathbb{R})}$ is, with the usual abuse of notation, the sheaf restricted to $X(\mathbb{R})$ and extended by zero outside $X(\mathbb{R})$.

We only need to prove the exactness on the stalks. For this we note that $\mathcal{F}_{|X(\mathbb{R})}$ is just the constant sheaf $\mathbb{Z}/2$ restricted to $X(\mathbb{R}) \subset Y$. In particular, if $x \in X(\mathbb{R})$, then $((1+S)\mathcal{F})_x = 0$. The exactness reduces here to $\mathcal{F}_x = (\mathcal{F}_{|X(\mathbb{R})})_x$ for $x \in X(\mathbb{R})$, which is trivial.

If $x \notin X(\mathbb{R})$ then $\mathcal{F}_x = F_1\oplus F_2$ ($F_i \cong \mathbb{Z}/2$) and $S(F_1) = F_2$. In particular, the kernel of $(1+S) : \mathcal{F}_x \longrightarrow \mathcal{F}_x$ is $(1+S)\mathcal{F}_x$. Since $(\mathcal{F}_{|X(\mathbb{R})})_x = 0$ in this case, this proves the exactness.

From the sequence (3.0) we get a long exact sequence of cohomology (<u>the Smith sequence</u>) :

$$(3.1) \quad \ldots \longrightarrow H^m(Y,(1+S)\mathcal{F}) \xrightarrow{\beta_m} H^m(Y,\mathcal{F}) \xrightarrow{\alpha_m}$$
$$H^m(Y,(1+S)\mathcal{F})\oplus H^m(Y,\mathcal{F}_{|X(\mathbb{R})}) \longrightarrow \ldots$$

Since G is finite, the Leray spectral sequence degenerates and we have $H^m(Y,\mathcal{F}) = H^m(X(\mathbb{C}),\mathbb{Z}/2)$ (see [Fl$_2$] p. 35). Hence (3.1) yields :

$$\ldots \longrightarrow H^m(Y,(1+S)\mathcal{F}) \xrightarrow{\beta_m} H^m(X(\mathbb{C}),\mathbb{Z}/2) \xrightarrow{\alpha_m}$$
$$H^m(Y,(1+S)\mathcal{F}) \oplus H^m(X(\mathbb{R}),\mathbb{Z}/2) \longrightarrow$$

If we denote $a_m = \dim(\mathrm{Ker}\ \alpha_m)$, we have :

(3.2) $$\Sigma \dim H^m(X(\mathbb{R}),\mathbb{Z}/2) = \Sigma\ (\dim H^p(X(\mathbb{C}),\mathbb{Z}/2) - 2a_p) \quad .$$

In general the a_p's introduced here are difficult to compute, but the following lemma will allow us to reduce the problem to something easier to compute :

(3.3) <u>Lemma</u> : <u>With the above notations we have</u> :

$$\mathrm{Ker}\ \alpha_m \supset (1+S)H^m(Y,\mathcal{F}) = (1+S)H^m(X(\mathbb{C}),\mathbb{Z}/2) \ .$$

<u>Proof</u> : Let γ be an m-cocycle representing $[\gamma] \in H^m(Y,\mathcal{F})$. Then $\gamma+\gamma^S$ represents the class $(1+S)[\gamma]$ and we have :

$$(1+S)(\gamma+\gamma^S) = 2\gamma + 2\gamma^S$$

and $$r(\gamma) = r(\gamma^S) \quad \text{or} \quad r(\gamma+\gamma^S) = 2r(\gamma) \ .$$

The fact that all the spaces in the exact sequence (3.1) are $\mathbb{Z}/2$-vector spaces shows that $\alpha_m(1+S)[\gamma] = 0$. This proves the lemma.

Writing $\lambda_m = \dim(1+S)H^m(X(\mathbb{C}),\mathbb{Z}/2)$, we can reformulate the lemma by saying :

(3.4) $$\lambda_m \leqslant a_m \quad .$$

Before giving our results in precise form we want to recall briefly a few well known facts of Galois cohomology.

(3.5) <u>Lemma</u> : <u>Let</u> A <u>be a free</u> $\mathbb{Z}$<u>-module of rank</u> n <u>and let</u> S <u>be an involution on</u> A. <u>Then there exists a basis</u> $(\alpha_1,\ldots,\alpha_b,\beta_1,\ldots,\beta_\lambda,\gamma_{\lambda+1},\ldots,\gamma_{n-b})$ <u>such that</u> :

<u>for</u> $1 \leqslant i \leqslant b$ $\qquad S(\alpha_i) = \alpha_i$,

<u>for</u> $1 \leq j \leq \lambda$ $\qquad S(\beta_j) = \alpha_j - \beta_j$,
<u>for</u> $\lambda < k \leq n-b$ $\qquad S(\gamma_k) = -\gamma_k$.

<u>Proof</u> : This is well known but for the sake of completeness we give a sketch of proof.

First note that Ker(1-S) ($=A^G$) is a direct factor in A. Secondly, that because $(1-S)\circ(1+S) = 0$, the morphism induced by (1+S) on A/Ker(1-S) is identically 0. In other words, the map induced by S on A/Ker(1-S) is -Id.

Combining these two facts we see that if we complete a basis of Ker(1-S), the matrix of S in such a basis will be of the form :

$$\begin{pmatrix} 1_b & N \\ 0 & -1_{n-b} \end{pmatrix}.$$

Using matrices of the form $\begin{pmatrix} 1 & B \\ 0 & C \end{pmatrix}$ to do base changes it is easy to see that we can : i) reduce N mod.2 ; ii) replace N by matrices of the form NC, C invertible. The lemma follows from this.

(3.5) is often formulated in the following form :

(3.5.1) <u>Let</u> A <u>and</u> S <u>be as in</u> (3.5). <u>Then</u> A <u>decomposes into a direct sum</u> $A_1 \oplus A_2 \oplus B_1 \oplus \ldots \oplus B_\lambda$ <u>where</u> $S_{|A_1} = Id$, $S_{|A_2} = -Id$ <u>and</u> $S_{|B_i}$ <u>has a matrix of the form</u> $\begin{pmatrix} 0 & 1 \\ 1 & 0 \end{pmatrix}$.

<u>Proof</u> : Take the basis $(\alpha_j - \beta_j, \beta_j)$ on the sub-module generated by (α_j, β_j) and reorder the terms.

b and λ are invariants for the action of S on A. We will say that λ is the <u>Comessatti characteristic</u> of (A,S). We will also say that (b,λ) are <u>invariants of</u> S. We can characterize b and λ in terms of the Galois cohomology of A under the action of $G = \{1,S\}$.

For b we have :

(3.6) $\qquad b = \text{rank}(A^G)$.

For λ first recall that :

(3.6.1) $H^1(G,A) = \mathrm{Ker}(1+S)/\mathrm{Im}(1-S)$

and $H^2(G,A) = \mathrm{Ker}(1-S)/\mathrm{Im}(1+S) = A^G/\mathrm{Im}(1+S)$,

both these groups being $\mathbb{Z}/2$-vector spaces.

Using (3.5) or (3.5.1) it is easy to see that :

(3.7) $\dim H^1(G,A) = n-b-\lambda$

$\dim H^2(G,A) = b-\lambda$.

We can give another characterization of λ by considering the action of G on $A_2 = A \otimes \mathbb{Z}/2 = A/2A$. We have :

(3.8) $\lambda = \dim(1+S)A_2$.

More generally let B be a $\mathbb{Z}/2$-vector space of dim n and let $G = \{1,S\}$ act on B. Writing $\lambda = \dim((1+S)B)$, we have :

(3.9) $\dim B^G = n-\lambda$

and if $i \geq 1$, $\dim H^i(G,B) = n-2\lambda$.

If we take $B = H^m(X(\mathbb{C}),\mathbb{Z}/2)$, the λ_m of (3.4) is just the λ introduced above. Since by (3.9), we have

$$\dim H^1(G,H^m(X(\mathbb{C}),\mathbb{Z}/2)) = \dim H^m(X(\mathbb{C}),\mathbb{Z}/2) - 2\lambda_m,$$

we can reformulate (3.2) and (3.4) :

(3.10) <u>Theorem</u> (Harnack-Thom-Krasnov inequality) : <u>If</u> X <u>is a real variety then</u> :

$$\Sigma \dim H^m(X(\mathbb{R}),\mathbb{Z}/2) \leq \Sigma \dim H^1(G,H^p(X(\mathbb{C}),\mathbb{Z}/2)) .$$

In [Kr] Krasnov gives a quite different proof of this theorem. We give a sketch of Krasnov's proof in Appendix 1 (A_1).

The most interesting case is the case when, instead of the inequality of (3.10), we have an equality. For this, following Krasnov, we introduce a definiton :

(3.11) Définition : A real variety X will be said to be Galois-Maximal or a GM-variety if we have :

$$\Sigma \dim H^m(X(\mathbb{R}),\mathbb{Z}/2) = \Sigma \dim H^1(G,H^p(X(\mathbb{C}),\mathbb{Z}/2)) \ .$$

We will in the sequel give examples of GM-varieties (see III 3. and IV. 1.), but it is important to note that already for surfaces not all surfaces are Galois Maximal (see Chap. V §4).

We are going to refine the result of theorem (3.10). We will need an extra hypothesis, namely that the cohomology of $X(\mathbb{C})$ has no 2-torsion.

We use again the notations of (2.1), that is $B_i = \dim H^i(X(\mathbb{C}),\mathbb{Q})$ and $b_i = \dim(H^i(X(\mathbb{C}),\mathbb{Q})^G)$. The above hypotheses on X imply that we also have :
$B_i = \dim_{\mathbb{Z}/2}H^i(X(\mathbb{C}),\mathbb{Z}/2)$. We will write $\lambda_i = \dim_{\mathbb{Z}/2}(1+S)H^i(X(\mathbb{C}),\mathbb{Z}/2)$.

With these notations and the hypothesis on the torsion we can write :

(3.12)
$$\dim_{\mathbb{Z}/2}H^1(G,H^i(X(\mathbb{C}),\mathbb{Z})) = B_i-b_i-\lambda_i$$

$$\dim_{\mathbb{Z}/2}H^2(G,H^i(X(\mathbb{C}),\mathbb{Z})) = b_i-\lambda_i \ .$$

We then have :

(3.13) Theorem : If X is a smooth projective real algebraic variety such that the cohomology of X(C) is without 2-torsion, then :

$$\Sigma \dim H^{2i}(X(\mathbb{R}),\mathbb{Z}/2) \leqslant \Sigma \dim H^2(G,H^p(X(\mathbb{C}),\mathbb{Z}))$$

$$\Sigma \dim H^{2i+1}(X(\mathbb{R}),\mathbb{Z}/2) \leqslant \Sigma \dim H^1(G,H^p(X(\mathbb{C}),\mathbb{Z})).$$

<u>We have equality in both cases if</u> X <u>is a GM-variety</u>.

<u>Proof</u> : Using the notations given above we can, by (3.9), rewrite (3.10) in the following form :

$$\Sigma \dim H^m(X(\mathbb{R}),\mathbb{Z}/2) \leq \Sigma (B_p-2\lambda_p).$$

On the other hand, we have by (2.1) :

$$\chi(X(\mathbb{R})) = \Sigma (2b_{2i}-B_{2i}) .$$

Since by (2.5), $B_{2i+1} = 2b_{2i+1}$, we get by taking the sum of the last two inequalities :

$$\Sigma \dim H^{2i}(X(\mathbb{R}),\mathbb{Z}/2) \leq \Sigma (b_p-\lambda_p) .$$

By (3.12) this is nothing but the first of the two inequalities of (3.13).

Taking the difference we get :

$$\Sigma \dim H^{2i+1}(X(\mathbb{R}),\mathbb{Z}/2) \leq \Sigma (B_p-b_p-\lambda_p),$$

which again by (3.12) is the second inequality of (3.13).

The last assertion of (3.13) follows from the above proof and definition (3.11).

(3.14) <u>Remark</u> : i) According to Krasnov [Kr] p. 255, the two inequalities of (3.13) still hold when $H^*(X(\mathbb{C}),\mathbb{Z})$ has 2-torsion.
ii) Again we note that all the results of this § are still true when $X(\mathbb{C})$ is a Kähler manifold and S an antiholomorphic involution on $X(\mathbb{C})$.

4. Real structures, divisors, Picard group and Chern classes

We assume again here that X is smooth and projective. Let (X,S) be a real structure on X and X_0 the corresponding real model. Note that by the smoothness hypothesis we can identify Weil divisors and Cartier divisors.

Let $D = \Sigma n_i D_i$ (where the D_i's are codimension 1 sub-varieties of X) be a Weil devisor. We define $S(D) = D^S$ to be $\Sigma n_i S(D_i)$. Recalling (1.9.1), we see that if D considered as a Cartier divisor is defined by (U_i, f_i), then D^S is defined by (U_i, f_i^S). We can reformulate this by saying : If $\mathcal{O}(D)$ is the invertible sheaf associated to D, we have with the notations of (1.7) :

(4.1) $$\mathcal{O}(D^S) = (\mathcal{O}(D))^S .$$

In the same way if $\mathcal{L}$ is an invertible sheaf on X and $D(\mathcal{L})$ is the associated Weil divisor, we have :

(4.2) $$D(\mathcal{L}^S) = (D(\mathcal{L}))^S .$$

We note a few miscellaneous facts here :

(4.3) *If D is an S-invariant divisor, then there exists a basis of the* $\mathbb{C}$-*vector space* $\Gamma(\mathcal{O}(D))$ *formed of S-invariant functions (that is functions such that* $f = f^S$ *in the notations of* (1.2)).

(4.4.) **Lemma** : *If X is projective with a real structure* (X,S), *then there exists on* X *an ample divisor satisfying* $D = D^S$.

Proof : If D' is an ample divisor on X, then by classical ampleness criterions D'^S is again ample and so is $D = D' + D'^S$.

The next result is deeper :

(4.5) **Proposition** : *Let* X *be a smooth projective variety with a real*

<u>structure</u> (X,S) <u>such that</u> $X(\mathbb{R}) \neq \emptyset$. <u>If</u> $D \in Div(X)$ <u>is such that</u> $D \sim D^S$ (<u>linear equivalence</u>), <u>then there exists</u> D' <u>such that</u> $D' \sim D$ <u>and</u> $D' = D'^S$.

<u>Proof</u> : Let X_0 be a real model for (X,S). We can in fact assume that $X = X_0 \times_{\mathbb{R}} \mathbb{C}$ and S is defined by the Galois action on X of $G = Gal(\mathbb{C}|\mathbb{R})$. We have in this cas $Div(X_0) = Div(X)^G$ and we can naturally identify $Div(X_0)$ and the sub-group of elements of Div(X) invariant under G. In this context (4.5) is nothing but the assertion that :

$$\text{(4.6)} \qquad Pic(X)^G = Pic(X_0) ,$$

where the action of G on Pic(X) is induced by the action of G on Div(X).

But this follows immediatly from the fact that under the hypotheses of (4.5), the Picard functor $Pic_{X|\mathbb{R}}$ is representable (see Grothendieck [Gr_2] Exp. 232).

<u>Remark</u> : If we have $D \approx D^S$ (algebraic equivalence), it is not clear that there always exists a $D' \approx D$ such that $D' = D'^S$ and this even if $X(\mathbb{R}) \neq \emptyset$. One of the difficulties lies in the fact that $H^1(G,Pic^o(X)) = H^1(G,H^1(X(\mathbb{C}),\mathbb{Z}))$ is not in general reduced to zero. We have in this direction partial results for the case of abelian varieties (see IV §3.).

We want now to describe the action of S on the Chern class of a divisor. For this we consider the exact sequence of sheaves on $X(\mathbb{C})$:

$$\text{(4.7)} \qquad 0 \longrightarrow \mathbb{Z} \xrightarrow{\times i} \mathcal{O}_{X(\mathbb{C})} \xrightarrow{\exp 2\pi} \mathcal{O}^*_{X(\mathbb{C})} \longrightarrow 0 ,$$

where $\mathcal{O}^*_{X(\mathbb{C})}$ is the sheaf of invertible holomorphic functions. Because of the $i = \sqrt{-1}$ factor this sequence is not a sequence of G-sheaves. To turn it into a sequence of G-sheaves we must twist the action of S on the constant sheaf $\mathbb{Z}$ by composing its action with $n \longmapsto -n$ in the fibres.

(4.8) <u>We will call</u> $\mathbb{Z}(1)$ <u>this twisted</u> G-<u>sheaf</u>. <u>In the same way we will</u>

<u>write</u> $H^i(X(\mathbb{C}),\mathbb{Z})(1)$ <u>for the</u> G-<u>module obtained from</u> $H^i(X(\mathbb{C}),\mathbb{Z})$ <u>by replacing the action of</u> S <u>by</u> $a \mapsto -S(a)$. <u>More generally, if</u> A <u>is any abelian group on which</u> G <u>acts, we will denote</u> A(1) <u>the twisted</u> G-<u>structure obtained by replacing the action of</u> S <u>by that of</u> -S.

On the Čech cocycle level one can explicitly describe the action of G on $H^i(X(\mathbb{C}),\mathcal{F})$ (where, to simplify, $\mathcal{F}$ is, say, a sub-sheaf of the sheaf of meromorphic functions) :

$$\left(U_{p_0\ldots p_i}, f_{p_0\ldots p_i}\right) \longmapsto \left(S\left(U_{p_0\ldots p_i}\right), f^S_{p_0\ldots p_i}\right).$$

As a consequence we have in particular :

(4.9) $$H^i(X(\mathbb{C}),\mathbb{Z})(1) = H^i(X(\mathbb{C}),\mathbb{Z}(1)) .$$

From the exact sequence of G-sheaves

(4.10) $$0 \longrightarrow \mathbb{Z}(1) \longrightarrow \mathcal{O}_{X(\mathbb{C})} \xrightarrow{\exp 2\pi} \mathcal{O}^*_{X(\mathbb{C})} \longrightarrow 0 ,$$

we get the cohomology sequence :

(4.11) $$0 \longrightarrow H^1(X(\mathbb{C}),\mathbb{Z})(1) \longrightarrow H^1(X(\mathbb{C}),\mathcal{O}_{X(\mathbb{C})}) \longrightarrow \mathrm{Pic}(X) \longrightarrow H^2(X(\mathbb{C}),\mathbb{Z})(1) \longrightarrow$$

(4.1) or (4.2) show that the G-module structure given here to Pic(X) by identification with $H^1(X(\mathbb{C}),\mathcal{O}^*_{X(\mathbb{C})})$ is the same as the one, considered earlier, induced by the G-action on Div(X). On the other hand the same identification gives the Néron-Severi group NS(X) (that is Div(X) modulo algebraic equivalence, which is the same as homological equivalence - see [Gr & Ha] p. 462) the structure of a sub G-module of $H^2(X(\mathbb{C}),\mathbb{Z})(1)$. In other words :

(4.12) <u>Lemma</u> : <u>If</u> (X,S) <u>is a real structure on an algebraic variety and if for a divisor</u> D <u>we have</u> $D \approx S(D)$ <u>(algebraic equivalence), then we have for the first Chern class of</u> D, $c_1(D)$ <u>considered as an element of</u> $H^2(X(\mathbb{C}),\mathbb{Z})$:

$$(c_1(D))^S = -c_1(D)$$

<u>where</u> S <u>operates via</u> S^*.

In the same class of ideas we can make the following remark :

Let $X(\mathbb{C})$ be a complex variety, not necessarily defined over $\mathbb{R}$, and let $X^\sigma(\mathbb{C})$ be its conjugate variety in the sense of (1.1). Let D be a divisor on $X(\mathbb{C})$ defined by the data (U_i, f_i). To D we can make correspond the divisor :

(4.13) D^σ defined on $X^\sigma(\mathbb{C})$ by $(U_i, \bar{f}_i)$.

Let d be the Chern class of D in $H^2(X_{top}, \mathbb{Z})$ obtained from the cohomology sequence associated to (4.8).

Since we have, if $f = \exp 2\pi i g$, $\bar{f} = \exp 2\pi(-i)\bar{g}$, we note that :

(4.14) <u>The Chern class of</u> D^σ <u>obtained by means of the exact sequence corresponding to</u> (4.8) <u>for</u> $\bar{O}_X$ <u>is equal to</u> $-d$ <u>in</u> $H^2(X_{top}, \mathbb{Z})$.

(compare with [Mi & St] Lemma 14.9 p. 168).

<u>Bibliographical Notes</u> :

The results of §1 are classical and are just translations in our special context of a few results of general descent theory, but they are probably much older.

A first version of (2.1), for surfaces, is in Comessatti [Co_3]. The idea of (2.4) comes from Deligne [De_1]. Again results of §2 can be found elswhere.

Our treatment of the Smith sequence is inspired by Floyd [Fl_2]. (3.10) can be found in Krasnov [Kr] ; the proof given here in [Si_5]. Definition (3.11) and a different version of (3.13) is also in [Kr].

The results of §4 are again classical.

APPENDIX TO CHAPTER I

KRASNOV'S RESULTS ON GALOIS-MAXIMAL VARIETIES

We give here a brief revue of certain results of [Kr].

The starting point of [Kr] lies in the consideration of certain spectral sequences of Grothendieck ([Gr_1] §5) and the cohomology functors $H^m(X;G,\mathcal{A})$ and $\mathcal{H}^m(G,\mathcal{A})$ (we refer to [Gr_1] p. 195 and following for the formalism). The results from [Gr_1] we will need are :

Theorem $A_1.1$ ([Gr_1] théorème 5.2.1) : *Let $\mathcal{A}$ be a sheaf of abelian groups on X, G a group acting on X and in a compatible way on $\mathcal{A}$ and let $Y = X/G$ be the topological quotient. There exists two spectral sequences with respective initial terms :*

$$I_2^{p,q}(X;G,\mathcal{A}) = H^p(Y,\mathcal{H}^q(G,\mathcal{A}))$$

and

$$II_2^{p,q}(X;G,\mathcal{A}) = H^p(G,H^q(X,\mathcal{A})) ,$$

converging to $H^n(X;G,\mathcal{A})$.

Theorem $A_1.2$ ([Gr_1] théorème 5.3.1) : *Let $y \in Y$, $x \in \pi^{-1}(y)$, G a finite group and G_x the stabilisor of x. We then have :*

$$\mathcal{H}^i(G,\mathcal{A})_y = H^i(G_x,\mathcal{A}_x) .$$

Theorem $A_1.3$ ([Gr_1] p. 206) : *If G acts trivially on X and $\mathcal{A}$ and if moreover, $\mathcal{A}$ is a sheaf of k-vector spaces then :*

$$H^n(X;G,\mathcal{A}) = \bigoplus_{p+q=n} (H^p(G,k) \otimes_k H^q(X,\mathcal{A})) .$$

Applying $A_1.3$ to the real part $X(\mathbb{R})$ of a real variety and the constant sheaf $\mathbb{Z}/2$, we get, noting that for all $p \geqslant 0$, $H^p(G,\mathbb{Z}/2) = \mathbb{Z}/2$:

$$(A_1.4) \qquad H^n(X(\mathbb{R});G,\mathbb{Z}/2) = \bigoplus_{0\leq q\leq n} H^q(X(\mathbb{R}),\mathbb{Z}/2) .$$

<u>Proposition</u> $A_1.5$ ([Kr] p. 251) : <u>Let</u> $n = \dim_{\mathbb{C}} X(\mathbb{C})$ <u>and</u> $\mathcal{A}$ <u>be a</u> G-<u>sheaf of abelian groups on</u> $X(\mathbb{C})$. <u>We have for</u> $r \geq 2n+1$:

$$H^r(X(\mathbb{C});G,\mathcal{A}) = H^r(X(\mathbb{R});G,\mathcal{A}_{|X(\mathbb{R})}).$$

<u>Proof</u> : Consider the exact sequence of sheaves on $X(\mathbb{C})$:

$$0 \longrightarrow \mathcal{S} \longrightarrow \mathcal{A} \longrightarrow \mathcal{A}_{|X(\mathbb{R})} \longrightarrow 0 ,$$

where $\mathcal{A}_{|X(\mathbb{R})}$ is here considered as a sheaf on $X(\mathbb{C})$, extended by 0 outside $X(\mathbb{R})$ and $\mathcal{S}$ is the kernel of the restriction morphism.

From this sequence we get an exact sequence of cohomology :

$$\longrightarrow H^r(X(\mathbb{C});G;\mathcal{S}) \longrightarrow H^r(X(\mathbb{C});G,\mathcal{A}) \longrightarrow$$
$$\longrightarrow H^r(X(\mathbb{C});G,\mathcal{A}_{|X(\mathbb{R})}) \longrightarrow H^{r+1}(X(\mathbb{C});G,\mathcal{S}) \longrightarrow$$

To prove $A_1.5$ is is enough to show that for $r > 2n$, $H^r(X(\mathbb{C});G,\mathcal{S}) = 0$. For this, we consider the spectral sequence :

$$I_2^{p,q} = H^p(Y,\mathcal{H}^q(G,\mathcal{S})) \Longrightarrow H^{p+q}(X(\mathbb{C});G,\mathcal{S}).$$

We first note that if $p > 2n = \dim Y$, then $I_2^{p,0} = H^p(Y,\mathcal{H}^0(G,\mathcal{S})) = 0$. Now if $x \in X(\mathbb{R})$, $\mathcal{S}_x = 0$ and if $x \in X(\mathbb{C})\backslash X(\mathbb{R})$, $G_x = 0$, so by $A_1.2$ we have

$$\mathcal{H}^q(G,\mathcal{S}) = 0 \qquad \text{for } q > 0.$$

This proves $A_1.5$.

<u>Theorem</u> $A_1.6$ ([Kr]) theorem 2.3) :

$$\Sigma \dim H^m(X(\mathbb{R}),\mathbb{Z}/2) \leq \Sigma \dim H^1(G,H^p(X(\mathbb{C}),\mathbb{Z}/2))$$

<u>and we have equality if and only if the spectral sequence</u> $II(X(\mathbb{C});G;\mathbb{Z}/2)$

<u>is trivial</u>.

<u>Proof</u> : By $A_1.4$ and $A_1.5$ we have :

$$\Sigma \dim H^p(X(\mathbb{R}),\mathbb{Z}/2) = \dim H^{2n+1}(X(\mathbb{R});G,\mathbb{Z}/2) = \dim H^{2n+1}(X(\mathbb{C});G,\mathbb{Z}/2) .$$

On the other hand, the spectral sequence

$$II_2^{p,q} = H^p(G,H^q(X(\mathbb{C}),\mathbb{Z}/2) \Longrightarrow H^{p+q}(X(\mathbb{C});G,\mathbb{Z}/2)$$

yields the inequality

$$\dim H^{2n+1}(X(\mathbb{C});G,\mathbb{Z}/2) \leq \sum_{p+q=2n+1} \dim H^p(G,H^q(X(\mathbb{C});G,\mathbb{Z}/2))$$

which is an equality if the spectral sequence is trivial.

Recalling that since the $H^q(X(\mathbb{C}),\mathbb{Z}/2)$ are $\mathbb{Z}/2$-vector spaces, we have :

$$\text{for } p \geq 1 \qquad H^p(G,H^q(X(\mathbb{C}),\mathbb{Z}/2)) = H^1(G,H^q(X(\mathbb{C}),\mathbb{Z}/2)).$$

This proves the inequality of $A_1.6$ and the "if" part of the last assertion. Since this is the most interesting part, and the only one we will use, we refer for the "only if" part to [Kr].

<u>Theorem</u> $A_1.7$ ([Kr] p. 262) : <u>If</u> X <u>is a smooth projective real algebraic surface such that</u> $X(\mathbb{R}) \neq \emptyset$ <u>and</u> $H^1(X(\mathbb{C}),\mathbb{Z}/2) = 0$ <u>then</u> X <u>is a GM-variety</u>.

<u>Proof</u> : By $A_1.6$ we only need to prove that the spectral sequence $II(X(\mathbb{C});G,\mathbb{Z}/2)$ is trivial.

For this we note that, because of the dimension, the only cases to consider are $1 \leq q \leq 4$. Since $H^1(X(\mathbb{C}),\mathbb{Z}/2) = H^3(X(\mathbb{C}),\mathbb{Z}/2) = 0$, we have $d_2^{p,1} = d_2^{p,2} = d_2^{p,3} = d_2^{p,4} = 0$. For the same reason we have $d_r^{p,1} = d_r^{p,3}$

= 0. The remaining $d_4^{p,q}$ being zero for questions of degree, the only cases to consider are $d_3^{p,2}$ and $d_3^{p,4}$. These two cases are settled by :

__Lemma__ A_1.8 ([Kr] p. 252 and p. 261) : __If__ X __is smooth__, dim X = n __and__ $X(\mathbb{R}) \neq \emptyset$, __then for all__ r __the__ $d_r^{p,r-1}$ (__resp. the__ $d_r^{p,2n}$) __of the spectral sequence__ $II(X(\mathbb{C});G,\mathbb{Z}/2)$ __are zero__.

__Proof__ : Let $x \in X(\mathbb{R})$. We consider the inclusion $\{x\} \hookrightarrow X(\mathbb{C})$. This inclusion induces a morphism of spectral sequences $II^{p,q}(X(\mathbb{C})) \longrightarrow II^{p,q}(\{x\})$ It also induces an isomorphism of G-modules $H^0(X(\mathbb{C}),\mathbb{Z}/2) \xrightarrow{\sim} H^0(\{x\},\mathbb{Z}/2)$ and hence isomorphisms $II_2^{p,0}(X(\mathbb{C})) \xrightarrow{\sim} II_2^{p,0}(\{x\})$.

Since the spectral sequence for $\{x\}$ is trivial, we find, by induction on r : $d_r^{p-r,r-1} = 0$.

For $d_r^{p,2n}$ we consider the inclusion $X(\mathbb{C})\setminus\{x\} \hookrightarrow X(\mathbb{C})$. Since $X(\mathbb{C})$ is smooth of topological dimension 2n, we have $H^{2n}(X(\mathbb{C})\setminus\{x\}) = 0$ and for k < 2n, isomorphisms of G-modules : $H^k(X(\mathbb{C})) \xrightarrow{\sim} H^k(X(\mathbb{C})\setminus\{x\})$. From the morphism of spectral sequence we find, by induction on r , $d_r^{p,2n} = 0$.

II. GENERAL RESULTS ON REAL ALGEBRAIC SURFACES

1. Real structures and the cup-product form

Let X be a smooth projective (complex) surface with a real structure (X,S). Let D_1 and D_2 be two divisors on X. We can define the intersection index of D_1 and D_2 (see Mumford [Mu_1] p. 84) by :

$$(D_1.D_2) = \chi(\mathcal{O}_X) - \chi(\mathcal{O}(-D_1)) - \chi(\mathcal{O}(-D_2)) + \chi(\mathcal{O}(-D_1-D_2)) .$$

where for a sheaf $\mathcal{L}$ on X

$$\chi(\mathcal{L}) = \Sigma(-1)^i \dim H^i(X,\mathcal{L})$$

If we apply I.(1.9) and I.(4.1) to this construction we find :

$$(1.1) \qquad (D_1^S.D_2^S) = (D_1.D_2) \quad .$$

We can in fact generalize (1.1) to the case where $X(\mathbb{C})$ is a Kähler surface and S an anti-holomorphic involution (to simplify notations we will also call in this case $(X(\mathbb{C}),S)$ a real structure on $X(\mathbb{C})$). For this recall that we can analytically define $(D_1.D_2)$ as the image under the cup-product form of $(c_1(D_1),c_1(D_2))$ (where $c_1(D)$ is the first Chern class of D),

$$H^2(X(\mathbb{C}),\mathbb{Z}) \times H^2(X(\mathbb{C}),\mathbb{Z}) \longrightarrow H^4(X(\mathbb{C}),\mathbb{Z}) = \mathbb{Z}.$$

The problem here is that, as noted in I.§4, the correspondance between D and $c_1(D)$ is not compatible with Galois action.

To make it compatible we must consider $c_1(D)$ as an element of $H^2(X(\mathbb{C}),\mathbb{Z})(1)$ (see I.(4.8)). If we do this we can do a construction analogous to the one made by Tate in [Ta] p. 96, and consider the cup-product as :

$$H^2(X(\mathbb{C}),\mathbb{Z})(1) \times H^2(X(\mathbb{C}),\mathbb{Z})(1) \longrightarrow H^4(X(\mathbb{C}),\mathbb{Z})(2) ,$$

where (2) is twisting twice. To show the equivalent of (1.1), we note that on the one hand $H^4(X(\mathbb{C}),\mathbb{Z})(2) = H^4(X(\mathbb{C}),\mathbb{Z})$ as a Galois module (for the trivial reason that $(-1)^2 = 1$) and on the other, since (x_1,y_1,x_2,y_2) defines the same orientation as $(x_1,-y_1,x_2,-y_2)$, the action of S (or rather S^*) on $H^4(X(\mathbb{C}),\mathbb{Z})$ is trivial by I.(1.13).

The above proves more generally that if $X(\mathbb{C})$ is a Kähler surface with a real structure $(X(\mathbb{C}),S)$ and if c and c′ are elements of $H^2(X(\mathbb{C}),\mathbb{Z})$ then :

$$(S(c).S(c')) = (-S(c).-S(c')) = (c.c').$$

We will write Q for the cup-product form on $H^2(X(\mathbb{C}),\mathbb{Z})$ and Q^+ and Q^- the restrictions to $H^2(X(\mathbb{C}),\mathbb{Z})^G$ (invariant part under S) and $H^2(X(\mathbb{C}),\mathbb{Z})(1)^G$ (invariant part under -S).

We recall also the notations of I.§2 : $B_i = \dim H^i(X(\mathbb{C}),\mathbb{Q})$, $b_i = \dim(H^i(X(\mathbb{C}),\mathbb{Q})^G)$, $h^{p,q}(X(\mathbb{C})) = \dim H^{p,q}(X(\mathbb{C}))$ and $\lambda_i = \dim_{\mathbb{Z}/2}(1+S)H^i(X(\mathbb{C}),\mathbb{Z}/2)$.

(1.2) <u>Proposition</u> : <u>If</u> X <u>is a smooth projective surface</u> (<u>or more generally a Kähler surface</u>) <u>with a real structure</u> (X,S), <u>then we have for the signatures of</u> Q^+ <u>and</u> Q^- :

$$Sg(Q^+) = (h^{0,2}, b_2-h^{0,2})$$
$$Sg(Q^-) = (h^{0,2}+1, B_2-b_2-h^{0,2}-1) .$$

<u>We also have</u> $|\det Q^+| = |\det Q^-| \leq 2^{\lambda_2}$,

<u>and we have equality if the cohomology of</u> $X(\mathbb{C})$ <u>is without</u> 2-<u>torsion</u>.

<u>Proof</u> : We know by the Hodge index theorem that the signature of Q (or rather $Q_{\mathbb{R}}$) restricted to $H^{1,1}(X(\mathbb{C})) \cap H^2(X(\mathbb{C}),\mathbb{R})$ is $(1,h^{1,1}-1)$ (see [Gr & Ha] p. 126).

Since X is projective there exists an ample divisor D on X such that $D^S = D$ (cf. I.(4.4)), hence such that $c_1(D)^S = -c_1(D)$ by I.(4.12) (the Chern class taken here in $H^2(X(\mathbb{C}),\mathbb{R})$).

We can generalize this to the Kähler case. Let $ds^2 = \Sigma\, d\varphi_i \otimes d\bar{\varphi}_i$ be a Kähler metric on $X(\mathbb{C})$ and :

$$(1.3) \qquad \omega = \sqrt{-1}/2\ \Sigma\, d\varphi_i \wedge d\bar{\varphi}_i$$

the associated (1,1)-form. Consider on X_{top} the analytic structure $X^\sigma(\mathbb{C})$ defined by the sheaf $\bar{O}_{X(\mathbb{C})}$ of anti-holomorphic functions (cf. I.(1.1)). The corresponding Kähler metric on $X^\sigma(\mathbb{C})$ is $d\bar{s}^2 = \Sigma\, d\bar{\varphi}_i \otimes d\varphi_i$ whose associated (1,1)-form is :

$$(1.4) \qquad \sqrt{-1}/2\ \Sigma\, d\bar{\varphi}_i \wedge d\varphi_i = -\,\omega\ .$$

Since S establishes an isomorphism between X and X^σ, $-\omega^S$ is again a (1,1)-form associated to a Kähler metric on $X(\mathbb{C})$ and so is $\omega + (-\omega^S)$.

Let ω be as above. Using the De Rham isomorphism we can associate to ω a class $[\omega]$ in $H^{1,1} \cap H^2(X(\mathbb{C}),\mathbb{R})$. We call such a class a <u>Kähler class</u>.

We recall two facts :

i) for all Kähler classes $[\omega]$, $Q([\omega],[\omega]) > 0$ (by definition of a Kähler class) ;

ii) Q restricted to $H^{1,1} \cap H^2(X(\mathbb{C}),\mathbb{R})$ has exactly one positive eigenvalue (by the Hodge index theorem).

Applying this to $\omega + (-\omega^S)$ we see that Q^- restricted to $H^{1,1} \cap H^2(X,\mathbb{R})$ has at least one positive eigenvalue by i) and hence exactly one by ii).

In other words, the restriction of Q^- to $H^{1,1} \cap H^2(X(\mathbb{C}),\mathbb{R})$ has a signature of the form (1,k).

Since $H^2(X(\mathbb{C}),\mathbb{R}) = H^2(X(\mathbb{C}),\mathbb{R})^G \perp H^2(X(\mathbb{C}),\mathbb{R})(1)^G$ we find that the restriction of Q^+ to $H^2(X(\mathbb{C}),\mathbb{R})^G \cap H^{1,1}$ is negative definite.

To end the proof of the first assertions of (1.2), recall that by the Hodge index theorem the restriction of Q to $(H^{0,2} \oplus H^{2,0}) \cap H^2(X(\mathbb{C}),\mathbb{R})$ is positive definite and that by I.(2.4) :

$$\dim(H^{0,2}(X) \oplus H^{2,0}(X)) \cap H^2(X(\mathbb{C}),\mathbb{R})^G = h^{0,2}$$

(resp. $\quad \dim(H^{0,2}(X) \oplus H^{2,0}(X)) \cap H^2(X(\mathbb{C}),\mathbb{R})(1)^G = h^{0,2})$

and hence by definition of b_2 : $\dim(H^{1,1}(X) \cap H^2(X(\mathbb{C}),\mathbb{R})^G) = b_2 - h^{0,2}$

(resp. $\quad \dim(H^{1,1}(X) \cap H^2(X(\mathbb{C}),\mathbb{R})(1)^G) = B_2 - b_2 - h^{0,2})$.

Summarizing we have :

$$Sg(Q^+) = (h^{0,2}, b_2 - h^{0,2}),$$

$$Sg(Q^-) = (h^{0,2} + 1, B_2 - b_2 - h^{0,2} - 1),$$

as announced.

For the last assertion we write $H^2(X(\mathbb{C}),\mathbb{Z})_f$ for $H^2(X(\mathbb{C}),\mathbb{Z})$ mod. torsion and λ' for the Comessatti characteristic corresponding to the action of S.

By I.(3.5) we have :

$$[H^2(X(\mathbb{C}),\mathbb{Z})_f : H^2(X(\mathbb{C}),\mathbb{Z})_f^G \oplus H^2(X(\mathbb{C}),\mathbb{Z})_f(1)^G] = 2^{\lambda'} .$$

On the other hand, we have, as an easy consequence of (1.1) and I.(3.5) :

$$(1.5) \qquad (H^2(X(\mathbb{C}),\mathbb{Z})_f^G)^{\perp} = H^2(X(\mathbb{C}),\mathbb{Z})_f(1)^G.$$

We will need a classical result (see for example Wilson [Wi] lemma

3.14, p. 61).

(1.6) Lemma : Let M be a free $\mathbb{Z}$-module with a bilinear form Q, A a subgroup and $B = A^{\perp}$. Then the absolute value of the determinant of Q restricted to A (or B) is equal to the index of $A \oplus B$ in M.

From this lemma and the above results and also because by definition $\lambda' \leq \lambda_2$, we get :

$$(1.7) \qquad |\det Q^{+}| = |\det Q^{-}| = 2^{\lambda'} \leq 2^{\lambda_2} .$$

Noting that in the absence of 2-torsion $\lambda' = \lambda_2$, this ends the proof of (1.2).

2. The Rokhlin-Kharlamov-Gudkov-Krakhnov congruences for surfaces

Throughout X will be a smooth projective real surface.

Let :

$$h^*(X(\mathbb{R})) = \dim H^*(X(\mathbb{R}),\mathbb{Z}/2) = \Sigma \dim H^i(X(\mathbb{R}),\mathbb{Z}/2)$$

and $$\chi(X(\mathbb{R})) = \Sigma (-1)^i \dim H^i(X(\mathbb{R}),\mathbb{Z}/2) .$$

With this notation formula I.(3.2) becomes :

$$(2.0) \qquad h^*(X(\mathbb{R})) = \Sigma (\dim H^p(X(\mathbb{C}),\mathbb{Z}/2) - 2a_p) .$$

(2.1) Definition : We will say that a smooth and projective real surface is an (M-r)-surface if $\Sigma a_i = r$ (the a_i's as in I.(3.2)).

(2.2) Proposition : If X is an M-surface (that is $\Sigma a_i = 0$) then :

$$2h^{0,2} - b_2 \equiv 0 \pmod{8} .$$

Proof : We first note that by I.(3.4), $\Sigma a_i \geq \Sigma \lambda_i$. This implies in particular that $\lambda_2 = \dim(1+S)H^2(X(\mathbb{C}),\mathbb{Z}/2) = 0$.

(1.2) (or (1.7)) then implies that $|\det Q^+| = 1$, or in other words, that Q^+ is unimodular.

We will need :

(2.3) <u>Lemma</u> : <u>If</u> X <u>is a smooth projective real surface</u>, <u>the restriction</u> Q^+ <u>of the cup-product form</u> Q <u>to</u> $H^2(X(\mathbb{C}),\mathbb{Z})^G$ <u>is even</u> (<u>of type</u> II).

<u>Proof</u> : Call w_k the k-th Stiefel-Whitney class of the Tangent bundle to $X(\mathbb{C})$ and v_2 the Wu-class in $H^2(X(\mathbb{C}),\mathbb{Z}/2)$.

An easy computation using Wu's theorem (see Milnor and Stasheff [Mi & St] p. 132) and the fact that $w_1 = 0$ ([Mi & St], p. 171) shows that $w_2 = v_2$.

On the other hand w_2 is the reduction mod. 2 (see [Mi & St] p. 171) of the first Chern class $c_1(T_X)$ of the complex tangent bundle which is equal to $-c_1(K)$, K a canonical divisor on X. By I.(4.12) we have $S^*(c_1(K)) = -c_1(K)$.

Combining all this we see that v_2 which verifies by definition :

$$\forall\ x \in H^2(X(\mathbb{C}),\mathbb{Z}/2) \qquad Q_{\mathbb{Z}/2}(x,x) = Q_{\mathbb{Z}/2}(x,v_2),$$

is the reduction mod. 2 of an element of $H^2(X(\mathbb{C}),\mathbb{Z})(1)^G$.

Since by (1.5) $H^2(X(\mathbb{C}),\mathbb{Z})^G$ and $H^2(X(\mathbb{C}),\mathbb{Z})(1)^G$ are orthogonal, this implies that for all $x \in H^2(X(\mathbb{C})\ \mathbb{Z})^G$, we have $Q(x,x) \equiv 0 \pmod{2}$, hence (2.3).

<u>End of proof of</u> (2.2) : Since Q^+ is unimodular and even, we know (see Serre [Se_2] p. 91) that its index is congruent to 0 mod. 8. But this index, by (1.2), is equal to $2h^{0,2}-b_2$.

(2.4) <u>Corollary</u> (Rokhlin's congruence) : <u>If</u> X <u>is a real</u> M-<u>surface</u>,

<u>then</u>

$$\chi(X(\mathbb{R})) \equiv \tau_Q \text{ (mod. 16)},$$

<u>where</u> τ_Q <u>is the index of</u> Q.

<u>Proof</u> : Using the fact that for surfaces $b_0 = b_4 = 1$ we can rewrite I.(2.1) :

(2.4.1) $$\chi(X(\mathbb{R})) = 2b_2 - B_2 + 2 .$$

On the other hand by the Hodge index theorem and the Hodge relations :

$$\begin{aligned}\tau_Q &= 2 + h^{0,2} + h^{2,0} - h^{1,1} \\ &= 2 + 4h^{0,2} - B_2 \quad .\end{aligned}$$

(2.4) then follows immediatly from (2.2).

(2.5) <u>Proposition</u> : (i) <u>If</u> X <u>is an</u> (M-1)-<u>surface</u>, <u>then</u> :

$$b_2 \equiv 2h^{0,2} \pm 1 \text{ (mod. 8)}$$

(ii) <u>If</u> $b_2 \equiv 2h^{0,2} \pm 3$ (mod. 8) <u>then</u> X <u>is at most an</u> (M-3)-<u>surface</u>.

For the proof we will need two lemmas.

(2.6) <u>Lemma</u> : <u>For a smooth projective real surface we have</u> $\Sigma a_i \equiv b_2$ (mod. 2).

<u>Proof</u> : We have

$$h^*(X(\mathbb{R})) + \chi(X(\mathbb{R})) \equiv 0 \text{ (mod. 4)} .$$

Using (2.0) and (2.4.1) we can reformulate this into :

$$\Sigma(B_i(\mathbb{Z}/2) - 2a_i) + 2b_2 - B_2 + 2 \equiv 0 \text{ (mod. 4)},$$

(where $B_i(\mathbb{Z}/2) = \dim H^i(X(\mathbb{C}),\mathbb{Z}/2)$) or, by Poincaré duality :

$$(2.7) \qquad 2(\Sigma a_i - b_2) \equiv 2B_1(\mathbb{Z}/2) + B_2(\mathbb{Z}/2) - B_2 \pmod{4}.$$

Now $2(B_1(\mathbb{Z}/2)-B_1) = B_2(\mathbb{Z}/2)-B_2$ by the invariance of the Euler characteristic and Poincaré duality and $B_1 \equiv 0 \pmod{2}$ by the Hodge relations. With this (2.7) becomes $4(B_1(\mathbb{Z}/2)) \equiv 2(\Sigma a_i - b_2) \pmod{4}$. Hence the lemma.

The next lemma is a direct consequence of Milgram's formula (see Milnor and Husemoller [Mi & Hu] p. 127 or [Gu] p. 59 and [Kh_1] for a direct proof).

(2.8) __Lemma__ : __Let__ A __be a free abelian group and__ Q __an even__ (__of type__ II) __bilinear form__. __Then, if__ $|\det Q| = 2$, __we have for the index__ τ_Q __of__ Q :

$$\tau_Q \equiv \pm 1 \pmod{8}.$$

__Proof of__ (2.5) : For (i). If $\Sigma a_i = 1$ we have by I.(3.4) and (1.2), $|\det Q^+| \leq 2$. Assume $|\det Q^+| = 1$. Then the same argument as the one used in the proof of (2.2) shows that we must have $b_2 \equiv 2h^{0,2} \pmod{8}$. But this implies $b_2 \equiv 0 \pmod{2}$, contradicting (2.2). Hence we must have $|\det Q^+| = 2$. Since Q^+ is even by (2.3), we can apply (2.8) and we find, by (1.2), $2h^{0,2}-b_2 \equiv \pm 1 \pmod{8}$ as announced.

For (ii) we note that if $b_2 \equiv 2h^{0,2} \pm 3 \pmod{8}$ then (2.2) and (i) above, imply that X is at most an (M-2)-surface. But if X were an M-2 surface, (2.6) would imply $b_2 \equiv 0 \pmod{2}$, hence the assertion.

(2.9) __Corollary__ (The Kharlamov-Gudkov-Krakhnov congruence) : __If__ X __is an__ (M-1)-__surface then__ :

$$\chi(X(\mathbb{R})) \equiv \tau_Q \pm 2 \pmod{16}.$$

__Proof__ : The proof is the same as the proof of (2.4), using (2.5)(i) in place of (2.2).

(2.10) <u>Remarks</u> : (i) The results of this § continue to hold if $X(\mathbb{C})$ is a Kähler surface with an antiholomorphic involution.
(ii) (2.4) and (2.9) where originally proved for varieties of even dimension using the Atiyah-Singer formula for the signature of an involution (see [Ro], [Kh_1] or [Gu]).

3. Bounds for the number of connected components and the $h^1(X(\mathbb{R}))$ of a real algebraic surface

We will use the notations of §2 and I.§2 and §3. We will also use the additional notation : $h^1(X(\mathbb{R})) = \dim H^1(X(\mathbb{R}),\mathbb{Z}/2)$.

We have :

$$(3.1) \qquad h^1(X(\mathbb{R})) = \frac{h^*(X(\mathbb{R})) - \chi(X(\mathbb{R}))}{2} .$$

Using the formulation given in §2 of the formulas (2.0) and (2.4.1) and the fact that by Poincaré duality $B_1(\mathbb{Z}/2) = B_3(\mathbb{Z}/2)$, we can rewrite (3.1) in the following form :

$$(3.2) \qquad h^1(X(\mathbb{R})) = \frac{2B_1(\mathbb{Z}/2) + B_2(\mathbb{Z}/2) + B_2 - 2(\Sigma a_i+b_2)}{2} .$$

If the cohomology of $X(\mathbb{C})$ is without 2-torsion we have the simpler form :

$$(3.3) \qquad h^1(X(\mathbb{R})) = B_1 + B_2 - (\Sigma a_i+b_2) .$$

In both cases the maximum values of $h^1(X(\mathbb{R}))$ for given Betti numbers of $X(\mathbb{C})$, corresponds to the minimum of Σa_i+b_2.

To simplify notations we are going to give bounds for $h^1(X(\mathbb{R}))$ in the case where the cohomology of $X(\mathbb{C})$ has no 2-torsion. When there is 2-torsion similar results can easily be proven using (3.2) in the place of (3.3).

(3.4) <u>Theorem</u> : <u>If</u> X <u>is a smooth projective real algebraic surface and</u>

<u>if</u> $H^*(X(\mathbb{C}),\mathbb{Z})$ <u>is without</u> 2-<u>torsion</u>, <u>then we have for the real part</u> $X(\mathbb{R})$.

(i) $h^1(X(\mathbb{R})) \leqslant B_1+B_2-h^{0,2}$ if $h^{0,2} \equiv 0 \pmod{8}$;

(ii) $h^1(X(\mathbb{R})) \leqslant B_1+B_2-h^{0,2}-1$ if $h^{0,2} \equiv \pm 1 \pmod{8}$;

(iii) $h^1(X(\mathbb{R})) \leqslant B_1+B_2-h^{0,2}-2$ if $h^{0,2} \equiv \pm 2$ or $4 \pmod{8}$;

(iv) $h^1(X(\mathbb{R})) \leqslant B_1+B_2-h^{0,2}-3$ if $h^{0,2} \equiv \pm 3 \pmod{8}$.

<u>Proof</u> : We are going to show how (iv) can be deduced from the results of §2 (the proof of the other inequalities is similar and in fact simpler).

We first recall I.(2.6), that is : $b_2 \geqslant h^{0,2}$. If $b_2 = h^{0,2}$, then, because $h^{0,2}$ ± 3 (mod.8), we have : $b_2-2h^{0,2} = -h^{0,2} \equiv \pm 3 \pmod{8}$. This implies by (2.5)(ii), that X is at most an (M-3)-surface. In other words, $\Sigma a_i \geqslant 3$. Since $b_2 = h^{0,2}$, (iv) follows from (3.3).

If $b_2 = h^{0,2}+1$, then $b_2 \equiv 2h^{0,2}+4$ or $b_2 \equiv 2h^{0,2}-2 \pmod{8}$. This implies by (2.2) and (2.5) (i), that X is at most an (M-2)-surface. Then $\Sigma a_i \geqslant 2$ and $\Sigma a_i+b_2 \geqslant 2+h^{0,2}+1$ and again (iv) follows from (3.3).

If $b_2 = h^{0,2}+2$ the same argument as above shows that X is at most an (M-1)-surface and the inequality (iv) is again verified. Finally if $b_2 \geqslant h^{0,2}+3$, (iv) is automatically verified.

<u>We write</u> $^{\#}X(\mathbb{R})$ <u>for the number of connected components of</u> $X(\mathbb{R})$. We have :

$$^{\#}X(\mathbb{R}) = \frac{h^*(X(\mathbb{R})) + \chi(X(\mathbb{R}))}{4} .$$

Using again the notations of §2 and of I.§2 and §3 we can reformulate this :

$$(3.5) \qquad ^{\#}X(\mathbb{R}) = \frac{4+2B_1(\mathbb{Z}/2)+B_2(\mathbb{Z}/2)-B_2+2b_2-2\Sigma a_i}{4} ,$$

or if the cohomology of $X(\mathbb{C})$ is without 2-torsion :

$$(3.6) \qquad {}^{\#}X(\mathbb{R}) = \frac{B_1 + b_2 - \Sigma a_i}{2} + 1 \quad .$$

To find an upper bound for ${}^{\#}X(\mathbb{R})$ it is here sufficient to bound $(b_2 - \Sigma a_i)$.

We start by searching an upper bound for b_2. By I.(2.7) we have :

$$b_2 \leqslant B_2 - h^{0,2} \ .$$

We are going to refine this bound. Let D be a divisor on X globally invariant under the action of S. We know by I.(4.12) that in this case $c_1(D)^S = -c_1(D)$. Hence $c_1(D) \in H^2(X(\mathbb{C}),\mathbb{Q})(1)^G$ (where $H^2(X(\mathbb{C}),\mathbb{Q})(1)^G$ is the invariant part of $H^2(X(\mathbb{C}),\mathbb{Q})$ under the action of $-S$).

We formulate this remark in more detail. The set of classes in $H^2(X(\mathbb{C}),\mathbb{Q})$ coming from divisors is the image of the Néron-Severi group NS(X) under the mapping :

$$\mathrm{Pic}(X) \longrightarrow H^2(X(\mathbb{C}),\mathbb{Z}) \longrightarrow H^2(X(\mathbb{C}),\mathbb{Q}) \ .$$

Now consider the G-module structure on NS(X) induced by the natural action of G on Div(X). Because of the twisting we have introduced in the exact sequence defining Pic(X) (see I.(4.11)), the image of $NS(X)^G$ under the above mapping lies in $H^2(X(\mathbb{C}),\mathbb{Q})(1)^G$. Since $H^2(X(\mathbb{C}),\mathbb{Q})^G$ and $H^2(X(\mathbb{C}),\mathbb{Q})(1)^G$ are orthogonal, we can improve the bound for b_2 given above :

$$(3.7) \qquad b_2 \leqslant B_2 - h^{0,2} - r \ ,$$

where $r = \mathrm{rank}\ NS(X)^G = \mathrm{rank}\ NS(X_o)$ (X_o a real model of (X,S)).

For projective surfaces (which are algebraic) we always have, because of I.(4.4), $r \geqslant 1$.

On the other hand, for a generic surface (or more precisely a surface generic in a family of surfaces for which $h^{0,2} > 0$), we have rank $NS(X) = 1$.

This justifies the fact that in practice we will use the bounds :

$$(3.8) \qquad h^{0,2} \leq b_2 \leq B_2-h^{0,2}-1 .$$

Note that if in the family of surfaces under consideration $h^{0,2} = 0$, we may not have rank $NS(X) = 1$ for a generic member (for example for cubic surfaces in $\mathbb{P}^2$ we have $r \geq 3$ - see Chap. VI (5.5)) and in this case (3.7) gives a better bound.

Using (3.6) and (3.8) and again (2.2) and (2.5) we obtain, reasoning exactly in the same way as in the proof of theorem (3.4) :

(3.9) <u>Theorem</u> : <u>If</u> X <u>is a smooth and projective real algebraic surface and if</u> $H^*(X(\mathbb{C}),\mathbb{Z})$ <u>is without torsion we have for the number of connected components</u> $^\#X(\mathbb{R})$ <u>of the real part</u> :

$$\text{(i)} \quad {}^\#X(\mathbb{R}) \leq \frac{B_1+B_2-h^{0,2}+1}{2} \text{ if } B_2-3h^{0,2}-1 = h^{1,1}-h^{0,2}-1 \equiv 0 \pmod{8}$$

$$\text{(ii)} \quad {}^\#X(\mathbb{R}) \leq \frac{B_1+B_2-h^{0,2}}{2} \text{ if } h^{1,1}-h^{0,2}-1 \equiv \pm 1 \pmod{8}$$

$$\text{(iii)} \quad {}^\#X(\mathbb{R}) \leq \frac{B_1+B_2-h^{0,2}-1}{2} \text{ if } h^{1,1}-h^{0,2}-1 \equiv \pm 2 \text{ or } 4 \pmod{8}$$

$$\text{(iv)} \quad {}^\#X(\mathbb{R}) \leq \frac{B_1+B_2-h^{0,2}-2}{2} \text{ if } h^{1,1}-h^{0,2}-1 \equiv \pm 3 \pmod{8}.$$

<u>Proof</u> : As said above, the proof of (3.9) is a replica of the proof given for (3.4). To see this, note that by (3.8), $b_2 \leq B_2-h^{0,2}-1$ and hence $b_2-2h^{0,2} \leq B_2-3h^{0,2}-1 = h^{1,1}-h^{0,2}-1$. Letting b_2 take its different possible values we can argue, using (2.2) and (2.5), as in the proof of (3.4).

Note here again, that using (3.5) in place of (3.6), it is possible to give a formulation of (3.9) valid in case $H^2(X(\mathbb{C}),\mathbb{Z})$ has 2-torsion.

For surfaces of degree n in $\mathbb{P}^3$ we can give a nicer formulation of (3.9). For such surfaces we know how to compute the Betti and Hodge numbers. We have :

$$B_1 = 0,\ B_2 = n^3-4n^2+6n-2,\ h^{0,2} = h^{2,0} = p_g = \frac{(n-1)(n-2)(n-3)}{6}$$

(see [Gr & Ha] p. 601-602). From this we see that :

$$B_2-3h^{0,2}-1 = n(n-1)^2/2 \quad \text{and} \quad B_2-h^{0,2} = (5n^3-18n^2+25n-6)/6.$$

On the other hand :

$$n(n-1)^2/2 \equiv \begin{cases} n/2 & (\text{mod.}8) \quad \text{if } n \text{ is even} \\ 0 & (\text{mod.}8) \quad \text{if } n \equiv 1 \ (\text{mod.}4) \\ -2 & (\text{mod.}8) \quad \text{if } n \equiv 3 \ (\text{mod.}4) \end{cases}$$

Hence :

(3.10) <u>Corollary</u> : <u>Let</u> X <u>be a smooth projective surface in</u> $\mathbb{P}^3$ <u>defined by a real irreducible polynomial of degree</u> n. <u>Then if</u> $P(n) = (5n^3-18n^2+25n)/12$ <u>the number of connected components of</u> $X(\mathbb{R})$ <u>is bounded by</u> :

(i) $\#X(\mathbb{R}) \leq P(n)$ if $n \equiv 0$ (mod.16) or $n \equiv 1$ (mod.4) ;
(ii) $\#X(\mathbb{R}) \leq P(n)-1/2$ if $n \equiv \pm 2$ (mod.16) ;
(iii) $\#X(\mathbb{R}) \leq P(n)-1$ if $n \equiv \pm 4$ or 8 (mod.16) or $n \equiv 3$ (mod.4);
(iv) $\#X(\mathbb{R}) \leq P(n)-3/2$ if $n \equiv \pm 6$ (mod.16) .

(3.11) <u>Remark</u> : (3.4) and the proof of (3.4) still hold when $X(\mathbb{C})$ is a Kähler surface and S an anti-holomorphic involution. (3.9) also holds, but not the proof. The reason for this is that we have used (3.7) which has no sense for Kähler surfaces. To prove (3.9) under this hypothesis we must use the fact that there exists a Kähler class $[\omega]$ satisfying $[\omega]^S = -[\omega]$ (see the proof of (1.2)) and that hence (3.8) holds.

4. Examples of real algebraic surfaces where $h^1(X(\mathbb{R}))$ or the number of connected components is maximum.

We are going to prove here that the bounds given in (3.4) and (3.9) are sharp, at least for small values of the Hodge numbers.

For (3.4) (i) : We have $\mathbb{P}^2$, certain cubics of $\mathbb{P}^3$ (see Chap.VI §5) and more generally all rational surfaces obtained by blowing up real points of $\mathbb{P}^2$ (cf. §6). In all these examples we have $h^{0,2} = p_g = 0$. It would be interesting to know if the bounds are still sharp when $h^{0,2} = 8,16,\ldots$. For other examples with $h^{0,2} = 0$ see the examples for (3.9) (ii) below.

For (3.4) (ii) : There is an example of Kharlamov of a smooth surface of degree 4 in $\mathbb{P}^3$ whose real part is a 10 holed torus T_{10} (see VIII or VII (3.3) for the construction of such a surface). A smooth quartic in $\mathbb{P}^3$ is a K3 surface ; hence we have $B_1 = 0$, $B_2 = 22$, $h^{0,2} = 1$ and $B_1+B_2-h^{0,2}-1 = 22-2 = 20$, so this surface realizes the maximum (for the complex invariants see [B&P&V] p. 239).

Another example is given by an abelian surface whose real part is formed of 4 tori (see Chap. IV) ; hence $h^1(X(\mathbb{R})) = 8$. We have (see Mumford [Mu_2] p. 4) : $B_1 = 4$, $B_2 = 6$ and $h^{0,2} = 1$, hence the assertion on the maximum by (3.4) (ii).

For (3.4) (iii) : The simplest example would be a surface of degree 5 in $\mathbb{P}^3$. For such a surface we would have $B_1 = 0$, $B_2 = 53$ and $h^{0,2} = 4$. The maximum for $h^1(X(\mathbb{R}))$ in such a case is 47. Such a surface would have 2 connected components, but I have not been able, up to now, to prove the existence of a surface of this type.

We are going to build examples for (3.4)(iii) and (iv) by another method. The method will be explained in detail in Chap. VII and the construction in VII (4.3) and (4.4). For the moment we content ourselves with the description of these examples. They are both elliptic surfaces. For the first we start with the elliptic fibration over $\mathbb{P}^1$ defined in $\mathbb{P}^2 \times \mathbb{P}^1$ ((x,y,z) homogeneous coordinates for $\mathbb{P}^2$ and (t,u) for $\mathbb{P}^1$) by :

$$(4.1) \qquad y^2z = x^3 + \left(\frac{t^3}{u(t-u)(t+u)}\right) xz^2 ,$$

and take for X the associated minimal regular model. We will prove in VII (4.3) that for this surface we have : $X(\mathbb{R})$ connected, $h^1(X(\mathbb{R})) = 30$ and for the complex part $B_1 = 0$, $B_2 = 34$ and $h^{0,2} = 2$. By (3.4)(iii) the maximum for $h^1(X(\mathbb{R}))$ is reached.

For (3.4)(iv) we apply the same method to the elliptic fibration :

$$(4.2) \qquad y^2z = x^3 + \left(\frac{u^4}{(u-t)(u+t)(u-2t)(u+2t)}\right) z^3 .$$

For the associated minimal regular model we have (see VII (4.4)) : $X(\mathbb{R})$ is connected, $h^1(X(\mathbb{R})) = 40$ and for the complex part, $B_1 = 0$, $B_2 = 46$ and $h^{0,2} = 3$. By (3.4)(iv) the maximum is reached.

For (3.9)(i) : We have at least the trivial example given by $\mathbb{P}^2$, if for the moment nothing more.

For (3.9)(ii) : Let C be a curve of genus g such that $C(\mathbb{R})$ has g+1 components (an M-curve) and let $X = \mathbb{P}^1 \times C$. We have $\#X(\mathbb{R}) = g+1$. Since X is a product we can apply the Künneth formula and we have $B_1 = 2g$ and $B_2 = 2$. Also, since this is a ruled surface, we have $h^{0,2} = 0$ (see [Ha] p. 371 or [Be] p. 49), $B_2-3h^{0,2}-1 = 1$ and $\frac{1}{2}(B_1+B_2-h^{0,2}) = g+1$ and the maximum is reached. Note that we also have $h^1(X(\mathbb{R})) = 2(g+1)$. Hence the maximum for (3.4)(i) is also reached.

For (3.9)(iii) : We have the example of an abelian surface with 4 connected components (4 T_1's). As noted before, we have for such a surface $B_1 = 4$, $B_2 = 6$, $h^{0,2} = 1$, $h^{1,1} = 4$. Hence we are in case (3.9)(iii) and the maximum is reached.

Another example is given by a K3 surface with 10 components (apparently Rohn [Rh] was the first to build such an example as a quartic in $\mathbb{P}^3$, see [Gu]). For such a surface we have $B_1 = 0$, $B_2 = 22$, $h^{0,2} = 1$ and $h^{1,1} = 20$ and it is easy to see that 10 is the maximum given by (3.9)(iii).

For (3.9)(iv) : Let E be a curve of genus 1 and C a curve of genus 2 such that $E(\mathbb{R})$ has 2 components and $C(\mathbb{R})$ has 3. Let $X = E\times C$. By Künneth's formula, we have $^{\#}X(\mathbb{R}) = 6$, $B_1 = 6$ and $B_2 = 10$. We are going to prove that $h^{0,2} = 2$. This will imply $h^{1,1} = 6$ and $h^{1,1}-h^{0,2}-1 = 3$. But in this case $\frac{1}{2}(B_1+B_2-h^{0,2}-2) = 6$ and the maximum will be reached.

To compute $h^{0,2} = \dim H^2(X,\mathcal{O}_X)$, we use Künneth's formula for $\mathcal{O}_X$. Since $H^2(E,\mathcal{O}_E) \cong H^2(C,\mathcal{O}_C) = 0$ we have :

$$H^2(X,\mathcal{O}_X) \cong H^1(E,\mathcal{O}_E) \otimes H^1(C,\mathcal{O}_C)$$

hence $h^{0,2} = 1\times 2 = 2$.

The examples we have given up to now are all of surfaces of special type (in the sense that none are of general type). We give here two examples of surfaces of general type, one for which the bound in (3.4) is reached and one for which the bound in (3.9) is reached.

We start by recalling some terminology about real plane curves (see [Gu]). The real part of smooth real curve in $\mathbb{P}^2$ of even degree consists of ovals (that is (topological) curves homeomorphic to S^1 that bound a disc in $\mathbb{P}^2(\mathbb{R})$). We will say that an oval α lies inside another oval β if it lies in the component of $\mathbb{P}^2(\mathbb{R})\setminus\beta$ homeomorphic to a disc. We will say that the curve is of type $\frac{k}{1}\,\ell$ if k ovals lie inside 1 oval (called the principal oval) and ℓ lie outside. For example fig. 1 below represents a curve of type $\frac{4}{1}\,3$:

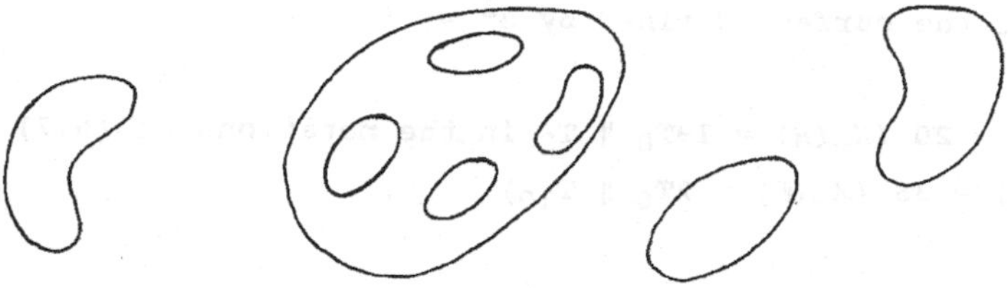

Fig. 1

Now by a method of Viro (see [Vi_1]), it is possible to build a smooth real curve of degree 8 (hence of genus 21) and of type $\frac{19}{1}\,2$

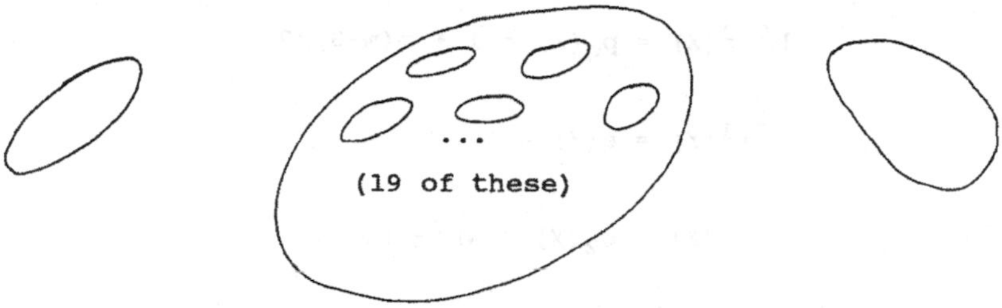

Choose an open affine covering $\{U_1,U_2,U_3\}$ of $\mathbb{P}^2$ and f_1, f_2, f_3 (f_i real in $\Gamma(U_i,\mathcal{O}_{\mathbb{P}^2}) = A_i$) defining C as a divisor on $\mathbb{P}^2$. We can always choose these such that the signs of f_i and f_j are the same on $U_i \cap U_j$. Define : $B_i = A_i[z]/z^2{=}f_i$ and $V_i = \operatorname{Spec} B_i$. The V_i's can be glued together to form a projective surface X defined over $\mathbb{R}$ (this because of the sign choice).

Since the curve C is non singular, X is non singular by construction and is a double covering of $\mathbb{P}^2$ ramified along the curve C.

To simplify notations we assume that the curve C is such that $C(\mathbb{R}) \subset \mathbb{R}^2$ and is defined by $f = 0$ in $\mathbb{R}^2$ (and in $\mathbb{C}^2$) ; then the part of $X(\mathbb{R})$ that lies in $\mathbb{R}^3$ is defined by $z^2 = f$.

In fact, depending on the choice of sign for f, we have two surfaces. Assume $f > 0$ at infinity, then we call X_+ the surface defined by

$z^2 = f$ and X_- the surface defined by $z^2 = -f$.

$^{\#}X_+(\mathbb{R}) = 20$ ($X_+(\mathbb{R}) = 19T_0 \amalg T_3$ in the notations of (6.7) below) and $h^1(X_-(\mathbb{R})) = 38$ ($X_-(\mathbb{R}) = 2T_0 \amalg T_{19}$).

The complex invariants of X_+ and X_- are the same, so we write X for either one. To compute these complex invariants we use general results on double coverings (see [B&P&V] p. 183) :
If X is a double covering of $\mathbb{P}^2$ ramified over a curve C of degree 2m having at most simple singularities then :

$$h^{0,2}(X) = p_g(X) = 1 + m(m-3)/2$$

$$h^{0,1}(X) = q(X) = 0$$

$$\chi(X) = c_2(X) = 4m^2 - 6m + 6 .$$

Since in our case m = 4 we find : $h^{0,2} = 3$, $B_1 = 2q = 0$, $B_2 = \chi(X)-2 = 44$. This implies that we are in cases (3.4)(iv) and (3.9)(iii) and the assertion on the maximum follows.

The bounds given in (3.10) are sharp for n = 1, 2 and 4 (recall the example of Rohn, given above, of a quartic in $\mathbb{P}^3$ with 10 components). For n = 3, the bound given by (3.10)(iii) is 3, but it is well known that a cubic surface in $\mathbb{P}^3$ has at most 2 components (see VI §5). It is however interesting to see that this is in fact a "false counter example". The difference between the bound given by (3.10) and the actual bound comes from the fact that in establishing (3.10) we have used (3.8) instead of (3.7) and that for cubics $r = \operatorname{rank} NS(X)^G \geqslant 3$ (see VI §5). If we had used (3.7) we would have found $B_2 - 3h^{0,2} - r = 7-3 = 4$. This means that we are in a case corresponding to (3.9)(iii) and in such a case we have $(B_2 - h^{0,2} - r)/2 = 4/2 = 2$, which is precisely the maximum found for cubics.

For n > 4 nothing much is known. The best examples found up to now in this direction are due to Viro [Vi_2]. We sketch here the principle of this construction.

Let C be a smooth M-curve of degree $m^2/2$ ($m \equiv 0 \pmod{2}$) in $\mathbb{P}^3$ complete intersection of two surfaces Y and Y′ of respective degrees m/2 and m. C has genus $\frac{m^2}{4}(m + m/2 - 4) + 1$ (see [Ha] p. 352).

If we consider Y and Y′ as divisors in $\mathbb{P}^3$, we have $2Y \sim Y'$. In other words, there exists a rational function f such that $2Y + (f) = Y'$.

Now consider $X_t = 2Y + (1+tf)$. By Bertini's theorem we can, for every $\epsilon > 0$, find $t \in [0,\epsilon]$, such that X_t is smooth (hence a smooth surface of degree m). On the other hand, if t is small enough, $X_t(\mathbb{R})$ is a double covering of $Y(\mathbb{R})$ ramified along $C(\mathbb{R})$. If $C(\mathbb{R})$ has a "good" arrangement of branches in $Y(\mathbb{R})$, then $X_t(\mathbb{R})$ will have many components (note that in fact, because C is an M-curve even a "bad" arrangement will produce quite a few components. If m = 8 for example, then C is of genus 129 and $C(\mathbb{R})$ has 130 components. Taking Y to be any quartic containing C and choosing the sign of f adequately, $X_t(\mathbb{R})$ will have between 60 and 140 components).

By a modification of a method of Brusotti, Viro constructs curves such that :

$$\#X_t(\mathbb{R}) = (m^3 - 2m^2 + 4)/4 \qquad \text{for } m \equiv 0 \pmod{2} .$$

If $m \equiv 2 \pmod{4}$, he uses another construction to produce M-curves on Y such that :

$$\#X_t(\mathbb{R}) = (7m^3 - 24m^2 + 32m)/24$$

5. <u>**Necessary conditions on a real surface for**</u> $X(\mathbb{R})$ <u>**to be empty**</u>.

We give here some conditions for $X(\mathbb{R}) = \emptyset$, but see III. (4.2). We first note that to establish formula I.(2.1) we have not used the hypothesis that $X(\mathbb{R})$ was non empty. This implies that formula (2.4.1) of this Chapter is still valid if $X(\mathbb{R}) = \emptyset$. In particular if $X(\mathbb{R}) = \emptyset$ we have :

$$(5.1) \qquad 2b_2 = B_2 - 2 .$$

This implies that if B_2 is odd then $X(\mathbb{R})$ is never empty. To obtain more precise conditions, we will assume that $H^1(X(\mathbb{C}),\mathbb{Z}/2) = 0$.

(5.2) Theorem : Let X be a smooth real projective surface such that $X(\mathbb{R}) = \emptyset$ and $H^1(X(\mathbb{C}),\mathbb{Z}/2) = 0$. We then have :

(i) $b_2 = \lambda_2 = (B_2-2)/2$;

(ii) The restriction Q^+ of the cup-product form Q to $H^2(X(\mathbb{C}),\mathbb{Z})^G$ is congruent to 0 mod. 2 (that is, $Q^+(\alpha,\beta) \equiv 0 \pmod{2}$ for all α and β).

Proof : We have already seen that under the above hypotheses $b_2 = (B_2-2)/2$. We note that $H^1(X(\mathbb{C}),\mathbb{Z}/2) = 0$ implies that $H^*(X(\mathbb{C}),\mathbb{Z})$ is without 2-torsion, hence by (1.2) $|\det Q^+| = 2^{\lambda_2}$. Since by definition $\lambda_2 \leq b_2 = \operatorname{rank} H^2(X(\mathbb{C}),\mathbb{Z})^G$, the equality $\lambda_2 = b_2$ will follow from (ii) (since $Q^+ \equiv 0 \pmod{2}$ and Q^+ non degenerate imply $|\det Q^+| \geq 2^{b_2}$).

We use again the notations of I§2 and §3, $Y = X(\mathbb{C})/G$ and $\pi : X(\mathbb{C}) \longrightarrow Y$.

To prove (ii), we first note that $\pi : X(\mathbb{C}) \longrightarrow Y$ is an unramified covering of order 2 and that this implies :

$$(5.3) \qquad Q(\pi^*(c),\pi^*(c')) = 2Q_Y(c,c')$$

where Q_Y is the cup-product form on Y (Recall that since $X(\mathbb{C})$ has topological dimension 4, S is orientation preserving and Y orientable).

To prove (ii), we only need, by (5.3), to prove :

(5.4) Lemma : If X is a real projective surface such that $H^1(X(\mathbb{C}),\mathbb{Z}/2) = 0$, then $X(\mathbb{R}) = \emptyset$ implies that $\pi^*H^2(Y,\mathbb{Z}) = H^2(X(\mathbb{C}),\mathbb{Z})^G$ (Y and π as above).

Proof : Let F be a constant sheaf on $X(\mathbb{C})$ and consider $\tilde{F} = \pi_*(F)$ on Y.

Since $X(\mathbb{R}) = \emptyset$ (that is S has no fixed points), $\tilde{F}$ is easy to describe : we have $\tilde{F}_y = F_x \oplus F_{S(x)}$ (where $(x,S(x)) = \pi^{-1}(y)$) with the obvious action $S : F_x \longrightarrow F_{S(x)}$.

From this description we get 3 things :

(i) $(1+S)\tilde{F}$ is isomorphic to the constant sheaf with fibre F_x (any x in $X(\mathbb{C})$). We write F_Y for this sheaf ;

(ii) The morphism $H^i(Y,(1+S)\tilde{F}) \longrightarrow H^i(Y,\tilde{F})$ corresponding to the inclusion $(1+S)\tilde{F} \hookrightarrow \tilde{F}$ is naturally identified with $\pi^* : H^i(Y,F_Y) \longrightarrow H^i(X(\mathbb{C}),F)$ (recall that, because G is finite, $H^i(X(\mathbb{C}),F) \cong H^i(Y,\tilde{F})$ canonically) ;

(iii) If $c \in H^i(X(\mathbb{C}),F)^G$, then $2c \in \pi^*(H^i(Y,F_Y))$ (to see this, use (ii) and the fact that if $c \in H^i(Y,\tilde{F})^G$ then $c + S(c) = 2c$).

Because of (iii) and since rank $H^2(X(\mathbb{C}),\mathbb{Z})^G = b_2$, (5.4) will follow from :

$$\text{(5.5)} \qquad \dim \pi^*(H^2(Y,\mathbb{Z})) \otimes \mathbb{Z}/2 = b_2$$

To prove (5.5) we are going to use (i), (ii) and (iii) above, applied to $\mathcal{F} = \pi_*(\mathbb{Z}/2)$ and the Smith sequence (see I.(3.1)). Since $X(\mathbb{R}) = \emptyset$ this reduces to :

(5.6)

$$\cdots \longrightarrow H^i(Y,\mathbb{Z}/2) \xrightarrow{\pi^*} H^i(X(\mathbb{C}),\mathbb{Z}/2) \longrightarrow H^i(Y,\mathbb{Z}/2) \longrightarrow H^{i+1}(Y,\mathbb{Z}/2) \longrightarrow \cdots$$

Recalling that $\dim H^i(X(\mathbb{C}),\mathbb{Z}/2) = B_i$ and that $X(\mathbb{C})$ and Y are orientable and connected, we get from (5.6), writing $h^i(Y)$ for $\dim H^i(Y,\mathbb{Z}/2)$,

$$\text{(5.7)} \qquad 2 + 2h^2(Y) = 4h^1(Y) + B_2 .$$

Assume $\dim \pi^*H^2(Y,\mathbb{Z}/2) = h^2(Y) - p$. In this case (5.6) and (5.7) yield :

$$h^1(Y) = (1+p)/2 \quad \text{and} \quad h^2(Y) = p + B_2/2.$$

In other words, $\dim \pi^*(H^2(Y,\mathbb{Z}/2)) = B_2/2 = b_2+1$.

We apply this to the universal coefficient diagram :

$$\begin{array}{ccccccccc} 0 & \longrightarrow & H^2(X(\mathbb{C}),\mathbb{Z})\otimes\mathbb{Z}/2 & \longrightarrow & H^2(X(\mathbb{C}),\mathbb{Z}/2) & \longrightarrow & 0 & \longrightarrow & 0 \\ & & \uparrow \pi^* & & \uparrow \pi^* & & \uparrow & & \\ 0 & \longrightarrow & H^2(Y,\mathbb{Z})\otimes\mathbb{Z}/2 & \longrightarrow & H^2(Y,\mathbb{Z}/2) & \longrightarrow & \mathrm{Tor}(H^3(Y,\mathbb{Z});\mathbb{Z}/2) & \longrightarrow & 0 \end{array}$$

We first note that, since $H^2(X(\mathbb{C}),\mathbb{Z})$ is 2-torsion free, the 2-torsion of $H^2(Y,\mathbb{Z})$ is killed by π^*. This, the above computations and the commutativity of the diagram show that to prove (5.5) we only need to prove that $\dim \mathrm{Tor}(H^3(Y,\mathbb{Z});\mathbb{Z}/2) = 1$.

To prove this last point, we note that since we are considering an unramified double covering and since $H_1(X(\mathbb{C}),\mathbb{Z}) = H^3(X(\mathbb{C}),\mathbb{Z})$ is of rank 0 and without 2-torsion, we have

$$\dim \mathrm{Tor}(H_1(Y,\mathbb{Z});\mathbb{Z}/2) = \dim \mathrm{Tor}(H^3(Y,\mathbb{Z});\mathbb{Z}/2) = 1.$$

This proves (5.5) and hence (5.4) and ends the proof of (5.2).

(5.8) <u>Remark</u> : The conditions of (5.2) are not sufficient for $X(\mathbb{R})$ to be empty. To see this, we have the example of quadrics in $\mathbb{P}^3$. In this case $H^2(X(\mathbb{C}),\mathbb{Z})$ is generated by two elements and it is easy to construct examples where $X(\mathbb{R}) \neq \emptyset$, but where these 2 elements are defined over $\mathbb{R}$ and hence have there classes in $H^2(X(\mathbb{C}),\mathbb{Z})(1)^G$. This implies that we have : $b_2 = 0 = (B_2-2)/2$, $\lambda_2 = 0$ and $Q^+ = 0$ (since $H^2(X(\mathbb{C}),\mathbb{Z})^G = 0$).

6. Real structure, birational transformations and blowing up

Let X_o be a surface over $\mathbb{R}$ (in the scheme sense). We will state results in this § for both real surfaces $X \cong X_o \times_{\mathbb{R}} \mathbb{C}$ (see I.(1.15)) and for surfaces over $\mathbb{R}$. We will always denote the latter by X_o, Y_o, ..., and as usual, S will denote the involution on X induced by Galois action.

(6.0) Definition. Let X and X′ be two real surfaces and let X_o and X_o' be the corresponding real models. We will say that $f : X' \longrightarrow X$ is a real birational transformation if f is induced by a birational transformation $f_o : X_o' \longrightarrow X_o$ (that is, a transformation induced itself, by an isomorphisme $\varphi : \mathbb{R}(X_o) \longrightarrow \mathbb{R}(X_o')$ between the function fields). If such a transformation exists we will say that both X and X′ and X_o and X_o' are real birationally equivalent.

Let X and X_o be as above and let $p : X \longrightarrow X_o$ be the canonical morphism. Let x be a closed point of X_o. We then have either $p^{-1}(x) = x$ or $p^{-1}(x) = \{x_1, x_2\}$, depending on the fact that $x \in X_o(\mathbb{R})$ or not.

(6.1) Proposition : Let notations be as above and let Y be the surface obtained by blowing up x (resp. x_1 and x_2) in the first case (resp. the second) and $\pi : Y \longrightarrow X$ the corresponding morphism. Then : Y has a natural real structure induced by that of X and π is defined over $\mathbb{R}$.

Proof : Recall the following description of a monoidal transformation with centre a point x : let (z_1, z_2) be local parameters at x defined in U, then $V = \pi^{-1}(U)$ has equation $tz_2 = uz_1$ in $U \times \mathbb{P}^1$, where (t,u) are homogeneous coordinates of $\mathbb{P}^1$.

Assume $x \in X(\mathbb{R})$, that (z_1, z_2) satisfy condition I.(1.11) and U is S-invariant.

We define S_Y in V by $S_Y(z_1, z_2; t, u) = (z_1^S, z_2^S; \bar{t}, \bar{u})$ (where ¯ denotes complex conjugation in $\mathbb{P}^1$). Because of condition I.(1.11) this involution coincides with the one induced on $Y \backslash \pi^{-1}(x)$ by identification with $X\backslash\{x\}$. This proves (6.1) in the first case.

In the second let x_1 and $x_2 = S(x_1)$ be conjugate points. Let (z_1,z_2) be local coordinates at x_1, satisfying I.(1.11) and defined on U_1, U_1 chosen such that $x_2 \notin U_1$. By hypothesis (z_1^S,z_2^S) are local coordinates at x_2 defined on $U_2 = S(U_1)$.

Let $\pi : Y \longrightarrow X$ be the surface obtained by blowing up x_1 and x_2. As before, we can define $V_1 = \pi^{-1}(U_1)$ in $U_1 \times \mathbb{P}^1$ by $tz_2 = uz_1$ and $V_2 = \pi^{-1}(U_2)$ by $tz_2^S = uz_1^S$. We define S_Y on $V_1 \cup V_2$ by $S_Y(z_1,z_2;t,u) = (z_1^S,z_2^S;\bar{t},\bar{u})$. Again this involution coincides by construction with the one induced by the identification

$X\backslash\{x_1,x_2\} = Y\backslash\{\pi^{-1}(x_1),\pi^{-1}(x_2)\}$.

We have a converse :

(6.2) <u>Proposition</u> (Manin) : <u>Let</u> X_0 <u>be a smooth surface over</u> $\mathbb{R}$, $X = X_0 \times_{\mathbb{R}} \mathbb{C}$ <u>and</u> $p : X \longrightarrow X_0$ <u>the canonical projection</u>. <u>Let</u> D <u>be a divisor on</u> X <u>such that</u> : $D = F$ <u>with</u> $F = S(F)$ <u>or</u> $D = F + S(F)$ <u>and</u> $(F.S(F)) = 0$ <u>where in both cases</u> F <u>is an exceptional curve of the first kind on</u> X (<u>that is</u> : $F \cong \mathbb{P}^1_{\mathbb{C}}$ <u>and</u> $(F.F) = -1$). <u>Then</u> D <u>can be</u> "<u>contracted over</u> $\mathbb{R}$", <u>that is, there exists a smooth surface</u> Y_0 <u>over</u> $\mathbb{R}$ <u>and a monoidal transformation</u> $\pi : X_0 \longrightarrow Y_0$ (<u>that is, such that the corresponding transformation</u> $\pi_{\mathbb{C}} : X \longrightarrow Y$ <u>over</u> $\mathbb{C}$ <u>is of one of the forms described in</u> (6.1)) <u>with centre a closed point</u> $y \in Y_0$ <u>such that</u> $p_*(D) = \pi^{-1}(y)$.

<u>Proof</u> : Let $D = F$ with $F = S(F)$. We can blow down D over $\mathbb{C}$. Let $\pi : X \longrightarrow Y$ be the resulting surface and map.

Since $S(D) = D$, the restriction of S to the open sub-variety $X' = X\backslash D$ is well defined and defines a Galois action on $Y\backslash\pi(D) \cong X'$. Letting $y = \pi(D)$, we can extend this action by $S_Y(y) = y$ and we have $\pi \circ S = S_Y \circ \pi$.
The result follows, in this case, from I.(1.4).

Let $D = F + S(F)$. As before we can blow down F over $\mathbb{C}$. Call F_1

the image of $S(F)$ under this contraction. Since $(F.S(F)) = 0$, we have $(F_1.F_1) = -1$ and we can again blow down F_1.

Using the fact that again $S(D) = D$ we can do a construction similar to the one made in the preceding case and build a real structure (Y,S_Y) on the resulting surface, such that $S_Y(\pi(F)) = \pi(S(F))$. The theorem follows by I.(1.4).

Note that the case $(F.S(F)) \neq 0$ can arise (we will see explicit examples in Chap. V) and that in this case the divisor $S(F) + F$ cannot be contracted.

Under the hypotheses of (6.2), we have since F is exceptional, $S(F) = F$ or $S(F) \cap F = \emptyset$. With this remark (6.1) and (6.2) lead to :

(6.3) Definition : Let X_0 be a smooth surface over $\mathbb{R}$ and $p : X = X_0 \times_{\mathbb{R}} \mathbb{C} \longrightarrow X_0$ the canonical projection. We will say that $\pi : Y_0 \longrightarrow X_0$ is a monoidal transformation if the corresponding transformation $\pi_{\mathbb{C}} : Y = Y_0 \times_{\mathbb{R}} \mathbb{C} \longrightarrow X$ is of one of the forms described in (6.1).

A curve F_0 on X_0 is said to be exceptional of the first kind if : (i) $p^{-1}(F_0) = F$ or (ii) $p^{-1}(F_0) = F + F^S$, F irreducible and exceptional of the first kind on X and $F = F^S$ in (i) or $F \cap F^S = \emptyset$ in (ii).

Our next result is in fact very classical, and we only include it here because it is in general only proved in the algebraically closed case (one exception being Manin [Ma_2]).

(6.4) Proposition : Let $f : X_0 \longrightarrow X_0'$ be a (real) birational transformation between two smooth, projective surfaces over $\mathbb{R}$, then it is possible to factor f into a finite sequence of monoidal transformations (see (6.3)) and their inverses.

Proof : The only thing to note here is that if $x \in X = X_0 \times_{\mathbb{R}} \mathbb{C}$ is a point of indeterminacy of $f_{\mathbb{C}}$, then so is $S(x)$ and if blowing up x removes the indeterminacy at x, blowing up $S(x)$ will remove the one at $S(x)$. With this in mind, it is easy to see that we can follow, practically word for word, the proof given in the algebraically closed case (see for example Hartshorne [Ha] theorem 5.5 p. 412).

Recall :

(6.5) Definition : A smooth projective surface X_0 over $\mathbb{R}$ is said to be relatively minimal if every birational $\mathbb{R}$-morphism $f : X_0 \longrightarrow X'_0$, where X'_0 is a smooth projective surface over $\mathbb{R}$, is an isomorphism. In this case, we will also say that $X = X_0 \times_{\mathbb{R}} \mathbb{C}$ is real relatively minimal.

With this definition we get an immediat, corollary to (6.4) (see Manin [Ma_2] theorem 21.8 p. 107) :

(6.6) Corollary : If X_0 is a smooth projective surface over $\mathbb{R}$, then X_0 is relatively minimal (or equivalently $X = X_0 \times_{\mathbb{R}} \mathbb{C}$ is real relatively minimal) if and only if one of the two following equivalent conditions is satisfied :
(i) For every exceptional curve of the first kind on $X = X_0 \times_{\mathbb{R}} \mathbb{C}$, we have $L^S \neq L$ and $L^S \cap L \neq \emptyset$;
(ii) There are no exceptional curves of the first kind (in the sense of (6.3)) on X_0.

(6.7) We introduce a notation : We will write T_g for the g holed torus ($T_o = S^2$) and U_h for the non-orientable surface (compact, connected and without boundary) of Euler characteristic 1-h (in particular $U_0 = \mathbb{P}^2(\mathbb{R})$).

Let $\pi : Y \longrightarrow X$ be a blowing up with centre a real point $x \in X(\mathbb{R})$. Let $L = \pi^{-1}(x) \cap Y(\mathbb{R})$. Since $\pi^{-1}(x)$ is isomorphic to $\mathbb{P}^1_{\mathbb{C}}$ and has a real point, $L = \mathbb{P}^1(\mathbb{R})$ and in particular is connected.

Let X_1 (resp. Y_1) be the connected component of $X(\mathbb{R})$ (resp. $Y(\mathbb{R})$) containing x (resp. L).

$Y(\mathbb{R})$ has the same number of connected components as $X(\mathbb{R})$ and each component Y_i of $Y(\mathbb{R})$, distinct from Y_1, is isomorphic (real analytically) to $\pi(Y_i) = X_i$.

To study Y_1, we consider the exact sequence of cohomology with compact support and coefficients in $A = \mathbb{Z}/2$ or $\mathbb{Q}$.

$$\cdots \longrightarrow H_c^p(Y_1-L,A) \longrightarrow H^p(Y_1,A) \longrightarrow H^p(L,A) \longrightarrow H_c^{p+1}(Y_1-L,A) \longrightarrow \cdots$$

We have :

(i) $H_c^0(Y_1-L,A) = H_c^0(X_1-x,A) = 0$;

(ii) $\forall\, i > 0$, $H_c^i(Y_1-L,A) = H_c^i(X_1-x,A) = H^i(X_1,A)$;

(iii) $H^i(L,A) = A$ for $i = 0$ or 1 and $H^j(L,A) = 0$ for $j \geqslant 2$

(see for example Massey [Mas]). In other words we have :

$$(6.8) \qquad H^1(Y_1,\mathbb{Z}/2) = H^1(X_1,\mathbb{Z}/2) \times (\mathbb{Z}/2)(\ell)$$

where (ℓ) is the class of L, and we see that if X_1 is orientable of type T_g , then Y_1 is non-orientable of type U_{2g}.

If X_1 is non-orientable, we consider cohomology with coefficients in $\mathbb{Q}$. From (ii) and (iii) we get :

$$H^2(X_1,\mathbb{Q}) \longrightarrow H^2(Y_1,\mathbb{Q}) \longrightarrow 0 .$$

This implies in particular that Y_1 is non-orientable. Using this and (6.8) we find that, if X_1 is of type U_h, then Y_1 is of type U_{h+1}.

Summarizing :

(6.9) <u>Proposition</u> : <u>Let</u> X <u>be a smooth projective real surface</u>.

<u>If</u> $\pi : Y \longrightarrow X$ <u>is a blowing up with centre a point</u> $x \in X(\mathbb{R})$, <u>then</u> : $Y(\mathbb{R})$ <u>and</u> $X(\mathbb{R})$ <u>have the same number of connected components and if</u> (X_i) <u>is the set of components of</u> $X(\mathbb{R})$ <u>the</u> $Y_i = \pi^{-1}(X_i) \cap Y(\mathbb{R})$ <u>are the components of</u> $Y(\mathbb{R})$ <u>and</u> :

$$Y_i \cong X_i \qquad\qquad \text{if } x \notin X_i$$

$$Y_i \cong \begin{cases} U_{2g} & \text{if } X_i \cong T_g \\ U_{h+1} & \text{if } X_i \cong U_h \end{cases} \qquad \text{if } x \in X_i$$

<u>If</u> Y <u>is obtained from</u> X <u>by blowing up a pair of complex conjugate points</u>, <u>then</u> : $X(\mathbb{R}) \cong Y(\mathbb{R})$.

(6.10) <u>Corollary</u> 1 : <u>Let</u> X <u>be as in</u> (6.9) <u>and assume</u> X <u>contains a line</u> L, $\mathbb{R}$-<u>invariant and with real points</u>. <u>If the self intersection</u> (L.L) <u>is odd and</u> ≥ -1, <u>then the connected component of</u> $X(\mathbb{R})$ <u>containing</u> $L(\mathbb{R})$ <u>is non-orientable</u>.

<u>Proof</u> : If (L.L) = 2n+1, we can blow up n+1 pairs of complex conjugate points of L, replacing X by Y with $Y(\mathbb{R}) \cong X(\mathbb{R})$. Y then has a real line with self-intersection -1 (see for example Hartshorne [Ha] V §3). The corollary follows from (6.2) and (6.9).

(6.11) <u>Corollary</u> 2 : <u>Let</u> X_o <u>be a smooth projective surface over</u> $\mathbb{R}$. <u>Then if all the connected components of</u> $X_o(\mathbb{R})$ <u>are orientable</u>, <u>there exists a relatively minimal model</u> Y_o <u>of</u> X_o <u>such that</u> $Y_o(\mathbb{R}) \cong X_o(\mathbb{R})$.

<u>Proof</u> : If X_o is not minimal, the only possible exceptional curves of the first kind on X_o are by (6.10) of the form (ii) of (6.3). By (6.4), we only need to blow down these and we can do this without changing the real part by (6.9).

(6.12) <u>Corollary</u> 3 : <u>If</u> X_o <u>and</u> X_o' <u>are</u> $\mathbb{R}$-<u>birationally equivalent surfaces</u> (<u>smooth and projective</u>), <u>then</u> $X_o(\mathbb{R})$ <u>and</u> $X_o'(\mathbb{R})$ <u>have the same number of connected components</u>. <u>In particular if</u> $X_o(\mathbb{R})$ <u>is not connected then</u> X_o <u>is not</u> $\mathbb{R}$-<u>birationally equivalent to</u> $\mathbb{P}^2_{\mathbb{R}}$.

<u>Proof</u> : We only need to apply (6.4) and (6.9).

<u>Remarks</u> :

(i) (6.10) is only here as an easy application of (6.9). In the next chapter we will largely improve this result (see III.(4.2)).

(ii) We will prove a converse of the last assertion of (6.12) in Chapter VI (VI.(6.4) and (6.5)).

Bibliographical Notes :

(1.2) has been used by many authors. At any rate (1.7) is proved in [Kh_2], and the first part of (1.2) is, under a different form, in Nikulin [Ni].

As noted before, (2.4) and (2.9) were proved for varieties of arbitrary dimension, (2.4) by Roklin [Ro], (2.9) by Kharlamov [Kh_1] and independently by Gudkov and Krakhnov [Gu & Kr]. The rest of §2 is adapted from [Si_4].

§3 and a good many examples of §4 are taken from [Si_4].
(5.2) was originally proved for rational surfaces in [Si_5].

III. ALGEBRAIC CYCLES ON REAL VARIETIES AND APPLICATION TO RATIONAL SURFACES.

1. A bound for the dimension of $H^1_{alg}(X(\mathbb{R}),\mathbb{Z}/2)$.

We continue to use the notations of I.(1.5). <u>We will always assume in this part that X is a smooth projective real variety and that</u> $X(\mathbb{R}) \neq \emptyset$.

We say that $W(\mathbb{R})$ is an algebraic sub-set of $X(\mathbb{R})$ if $W(\mathbb{R}) = X(\mathbb{R}) \cap W$ where W is an algebraic subvariety of X (note that since X is projective by definition, $W(\mathbb{R})$ is compact). We will say that $W(\mathbb{R})$ is irreducible if this is the case for W.

Let $\mathfrak{Z}$ be the free abelian group generated by irreducible algebraic subsets of codimension 1 in $X(\mathbb{R})$. We have a map :

(1.1) $$r : \mathfrak{Z} \longrightarrow H^1(X(\mathbb{R}),\mathbb{Z}/2) .$$

We give here a sketch of the construction. Let $c \in \mathfrak{Z}$ and let $\tilde{c}$ be its reduction mod.2. Then $\tilde{c}$ definies an (n-1)-cycle with coefficients in $\mathbb{Z}/2$. To $\tilde{c}$ we can associate its class [c] in $H_{n-1}(X(\mathbb{R}),\mathbb{Z}/2)$. We define r(c) to be the Poincaré dual of [c] in $H^1(X(\mathbb{R}),\mathbb{Z}/2)$. For more details on this construction see [B&C&R] p. 235 and p. 269.

<u>We define</u> $H^1_{alg}(X(\mathbb{R}),\mathbb{Z}/2)$ <u>to be the image of</u> r.

Let $c \in \mathfrak{Z}$ and define $\gamma(c)$ to be the complexification of c. Obviously γ defines a map :

(1.2) $$\gamma : \mathfrak{Z} \longrightarrow Div(X)^G \quad .$$

On the other hand, we define a map

$$(1.3) \qquad \alpha' : \mathrm{Div}(X)^G \longrightarrow \mathfrak{Z}$$

in the following way : let $D = \Sigma\, n_i D_i$ be the decomposition of D into irreducible components (irreducible over $\mathbb{C}$). If D_i is defined over $\mathbb{R}$ we define $\alpha'(D_i) = D_i \cap X(\mathbb{R})$ if this set is of codimension 1 in $X(\mathbb{R})$ and $\alpha'(D_i) = 0$ if the codimension is strictly greater than 1. If D_i is not defined over $\mathbb{R}$ we let $\alpha'(D_i) = 0$ (this makes sense because if D_i is not defined over $\mathbb{R}$, $\mathrm{codim}(D_i \cap X(\mathbb{R})) > 1$). Finally we define $\alpha'(D) = \Sigma\, n_i \alpha'(D_i)$.

Composing α' with the natural map $\mathfrak{Z} \longrightarrow \mathfrak{Z} \otimes \mathbb{Z}/2 = \mathfrak{Z}(\mathbb{Z}/2)$, we get a map :

$$(1.4) \qquad \alpha : \mathrm{Div}(X)^G \longrightarrow \mathfrak{Z}(\mathbb{Z}/2)$$

and composing further with the map $\mathfrak{Z}(\mathbb{Z}/2) \longrightarrow H^1_{alg}(X(\mathbb{R}),\mathbb{Z}/2)$, we get :

$$(1.5) \qquad \rho : \mathrm{Div}(X)^G \longrightarrow H^1_{alg}(X(\mathbb{R}),\mathbb{Z}/2) \quad .$$

If $c \in \mathfrak{Z}$ we note that we clearly have $\gamma(c) \cap X(\mathbb{R}) = c$. In other words, $\alpha' \circ \gamma = \mathrm{Id}$. By the definition of ρ this implies that we have the commutative diagram :

$$(1.6) \qquad \begin{array}{ccc} \mathfrak{Z} & & \\ {\scriptstyle\gamma}\big\downarrow & \searrow^{r} & \\ \mathrm{Div}(X)^G & \xrightarrow{\ \rho\ } & H^1_{alg}(X(\mathbb{R}),\mathbb{Z}/2) \end{array} \quad .$$

Now assume that for $c \in \mathfrak{Z}$ the divisor $\gamma(c) \in \mathrm{Div}(X)^G$ is principal. Since $\gamma(c)$ is S-invariant, we can by I.(4.3) find f real such that $\gamma(c) = (f)$. But in this case $\gamma(c) \cap X(\mathbb{R}) = (f) \cap X(\mathbb{R})$ ("divisor" of real zeros and poles of f) and $\gamma(c) \cap X(\mathbb{R})$ bounds the set $\{x \in X(\mathbb{R}) \ / \ f(x) > 0\}$. In other words, we have $r(c) = 0 \in H^1(X(\mathbb{R}),\mathbb{Z}/2)$ (in fact, we have the much stronger result that $c \in \mathfrak{Z}$ is the "divisor" of real zeros and poles of a rational function f if and only if r(c) =

0 - see Bröcker [Br]). This implies that we can factorize r further by (recall that by I.(4.5) the map $Div(X)^G \longrightarrow Pic(X)^G$ is onto)

$$(1.7)\qquad \begin{array}{ccc} \mathfrak{Z} & \xrightarrow{\quad r \quad} & \\ \downarrow & & \searrow \\ Pic(X)^G & \xrightarrow{\alpha^*} & H^1_{alg}(X(\mathbb{R}),\mathbb{Z}/2) \end{array}$$

We note that by construction the map α^* is surjective. We are going to prove that this last map α^* can itself be factorized by a map :

$$(1.8)\qquad \beta : H^2(G, Pic(X)) \longrightarrow H^1_{alg}(X(\mathbb{R}),\mathbb{Z}/2) \quad .$$

To prove this we note that since (see I.(3.6.1)) ,

$$H^2(G,Pic(X)) = Pic(X)^G/(1+S)Pic(X),$$

we only need to prove that :

$$(1.9)\qquad (1+S)Pic(X) \subset Ker\ \alpha^* \ .$$

For this, let $d+d^S \in (1+S)Pic(X)$ and let D be a representative of d in Div(X). Let $D = \Sigma\, n_iD_i$ be the decomposition of D into irreducible divisors and assume that for $1 \leq i \leq k$, D_i is defined over $\mathbb{R}$ (S-invariant) and that for $i > k$, D_i is not defined over $\mathbb{R}$ (not S-invariant).

We have $D^S = \sum_{i=1}^{k} n_iD_i + \sum_{i>k} n_iD_i^S$ and, by definition of α', $\alpha'(D + D^S) = 2\sum_{i=1}^{k} n_i\alpha'(D_i)$ or $\alpha(D + D^S) = 0$. Hence $\alpha^*(d + d^S) = 0$. This proves (1.9) and the assertion about β.

(1.10) <u>Theorem</u> : <u>If</u> X <u>is a smooth</u>, <u>irreducible and projective real variety we have</u> :

$$\dim H^1_{alg}(X(\mathbb{R}),\mathbb{Z}/2) \leq \dim H^2(G,Pic(X))$$

$$\leq \dim H^1(G,H^1(X(\mathbb{C}),\mathbb{Z})) + \dim H^2(G,NS(X)).$$

Proof : The first inequality is a direct consequence of the surjectivity of β, which follows from the surjectivity of α^* and the definition of β.

To prove the inequality

$$\dim H^2(G,Pic(X)) \leq \dim H^1(G,H^1(X(\mathbb{C}),\mathbb{Z})) + \dim H^2(G,NS(X))$$

we consider the exact sequence of G-modules :

$$0 \longrightarrow Pic^o(X) \longrightarrow Pic(X) \longrightarrow NS(X) \longrightarrow 0 .$$

Applying Galois cohomology to this sequence, we see that

$$\dim H^2(G,Pic(X)) \leq \dim H^2(G,Pic^o(X)) + \dim H^2(G,NS(X))$$

To end the proof we only need to prove that :

$$(1.11) \qquad H^1(G,H^1(X(\mathbb{C}),\mathbb{Z})) = H^2(G,Pic^o(X)) \quad .$$

For this we consider the exact sequence of G-modules of I.(4.11) :

$$0 \longrightarrow H^1(X(\mathbb{C}),\mathbb{Z})(1) \longrightarrow H^1(X(\mathbb{C}),\mathcal{O}_{X(\mathbb{C})}) \longrightarrow Pic(X)$$
$$\longrightarrow H^2(X(\mathbb{C}),\mathbb{Z})(1) \longrightarrow \dots$$

Inserting $Pic^o(X)$ in this sequence we have :

$$0 \longrightarrow H^1(X(\mathbb{C}),\mathbb{Z})(1) \longrightarrow H^1(X(\mathbb{C}),\mathcal{O}_{X(\mathbb{C})}) \longrightarrow Pic^o(X) \longrightarrow 0.$$

We note that, since $H^1(X(\mathbb{C}),\mathcal{O}_{X(\mathbb{C})})$ is a $\mathbb{C}$-vector space, it is easy to prove that $H^i(G,H^1(X(\mathbb{C}),\mathcal{O}_{X(\mathbb{C})})) = 0$ for all $i > 0$ (recall I.(3.6.1)). With this remark, Galois cohomology applied to the last sequence yields the isomorphism $H^1(G,H^1(X(\mathbb{C}),\mathbb{Z})(1)) = H^2(G,Pic^o(X))$ (recall that since $G = \mathbb{Z}/2$, $H^{2n+1}(G,A) = H^1(G,A)$). But since $b_1 = B_1/2$ (I.(2.5)), this proves (1.11) by I.(3.7).

2. <u>A lower bound for</u> $H^1_{alg}(X(\mathbb{R}),\mathbb{Z}/2)$ <u>and some computations</u>.

We start with a consequence of (1.10) :

(2.1) <u>Lemma</u> : <u>If</u> X <u>is a GM-surface</u>, $X(\mathbb{R}) \neq \emptyset$, <u>and</u> $H^*(X(\mathbb{C}),\mathbb{Z})$ <u>is without 2-torsion</u>, <u>then</u> :

$$H^1_{alg}(X(\mathbb{R}),\mathbb{Z}/2) \neq H^1(X(\mathbb{R}),\mathbb{Z}/2)$$

<u>if one of the following two conditions holds</u>

- $H^1(G,H^1(X(\mathbb{C}),\mathbb{Z})) \neq 0$;
- $H^2(G,NS(X)) \neq H^2(G,H^2(X(\mathbb{C}),\mathbb{Z})(1)) = H^1(G,H^2(X(\mathbb{C}),\mathbb{Z}))$.

<u>Proof</u> : If X is a GM-surface we have by I.(3.13) :

$$\dim H^1(X(\mathbb{R}),\mathbb{Z}/2) = \Sigma \dim H^1(G,H^p(X(\mathbb{C}),\mathbb{Z}))$$

or using Poincaré duality, the fact that there is no 2-torsion and the fact that G acts trivially on H^0 and H^4 :

(2.2.) $\dim H^1(X(\mathbb{R}),\mathbb{Z}/2) = 2 \dim H^1(G,H^1(X(\mathbb{C}),\mathbb{Z})) + \dim H^1(G,H^2(X(\mathbb{C}),\mathbb{Z}))$.

Since in any case $NS(X) \subset H^2(X(\mathbb{C}),\mathbb{Z})(1)$ (see I.(4.11)) and by definition

$$H^2(G,H^2(X(\mathbb{C}),\mathbb{Z})(1)) \cong H^1(G,H^2(X(\mathbb{C}),\mathbb{Z})),$$

(2.1) follows from (2.2) and (1.10).

(2.3) <u>Remark</u> : (2.1) implies that the inequality

$$H^1_{alg}(X(\mathbb{R}),\mathbb{Z}/2) \neq H^1(X(\mathbb{R}),\mathbb{Z}/2)$$

is already for surfaces quite frequent and in fact, as we will see below, generic.

(2.4) <u>Proposition</u> (Bochnak-Kucharz-Shiota) : <u>Let</u> X <u>be a smooth projective real surface</u>. <u>If</u> X(ℝ) <u>has a non orientable component, then</u> $\dim H^1_{alg}(X(\mathbb{R}),\mathbb{Z}/2) \geq 1$.

<u>Proof</u> : We first recall that if X(ℝ) is non orientable then there exists $c \in H^1(X(\mathbb{R}),\mathbb{Z}/2)$ such that $\langle c.c\rangle = 1$, where $\langle.\rangle$ is cup product mod 2 on X(ℝ) (see Milnor and Husemoller [Mi&Hu] lemma (1.1) p. 101).

Let $v_1(\mathbb{R}) \in H^1(X(\mathbb{R}),\mathbb{Z}/2)$ be the Wu class. By Wu's theorem ([Mi&St] p. 132), $v_1(\mathbb{R}) = w_1(\mathbb{R})$, the Stiefel-Whitney class of the tangent bundle to X(ℝ). Also, by definition of the Wu class, we have

$$1 = \langle c.c\rangle = \langle c.w_1(\mathbb{R})\rangle .$$

In particular $w_1(\mathbb{R}) \neq 0$ in $H^1(X(\mathbb{R}),\mathbb{Z}/2)$.

On the other and, it is fairly obvious that $w_1(\mathbb{R})$ is an algebraic class. In fact, one can be even more precise, and prove that $w_1(\mathbb{R}) = \alpha^*(c_1(K))$, where $c_1(K) \in \mathrm{Pic}(X)^G$ is the canonical class on X. To see this, recall that $-c_1(K) = c_1(T_{X(\mathbb{C})})$, the first Chern class of the complex tangent bundle to X(ℂ). If we write $T_{X(\mathbb{C})}|_{X(\mathbb{R})}$ for the restriction of $T_{X(\mathbb{C})}$ to X(ℝ), then obviously $T_{X(\mathbb{C})}|_{X(\mathbb{R})} = T_{X(\mathbb{R})}\otimes_{\mathbb{R}}\mathbb{C}$. This proves by definition of α^*, that $w_1(\mathbb{R}) = \alpha^*(c_1(K))$.

In [B&C&R] p.239, Bochnak Coste and Roy prove that for any compact connected topological surface without boundary $\mathcal{Y}$ different from T_0, U_0 and U_1 (see II.(6.7)) there exists an algebraic surface X such that X(ℝ) is homeomorphic to $\mathcal{Y}$ and $H^1(X(\mathbb{R}),\mathbb{Z}/2) \neq H^1_{alg}X(\mathbb{R}),\mathbb{Z}/2)$.

On the other hand, we see as a direct consequence of (2.4) that if $X(\mathbb{R}) \cong U_0$ the projective plan $\mathbb{P}^2(\mathbb{R})$, then $H^1(X(\mathbb{R}),\mathbb{Z}/2) = H^1_{alg}(X(\mathbb{R}),\mathbb{Z}/2)$. Also, if $X(\mathbb{R}) \cong T_0$, that is the sphere, then we obviously have equality.

As noted in [B&C&R] this leaves the question open for the Klein Bottle U_1. We are going to prove that also in this case one can have inequality.

The proof we are going to give can easily be generalized to prove the result of [B&C&R] quoted above, with the additional information that the inequality $H^1_{alg}(X(\mathbb{R}),\mathbb{Z}/2) \neq H^1(X(\mathbb{R}),\mathbb{Z}/2)$ is in fact generic for algebraic surfaces in $\mathbb{P}^3$ (and in fact for algebraic surfaces in general).

Our example is a quintic surface in $\mathbb{P}^3$. We start with a smooth curve C of degree 5 in $\mathbb{P}^2$ such that X(ℝ) has 2 components. Most any curve of this type will do, but to fix ideas we choose a specific one defined by :

$$P(x,y,z) = x^5 + xyz^3 + y^5 - z^5/32 = 0 .$$

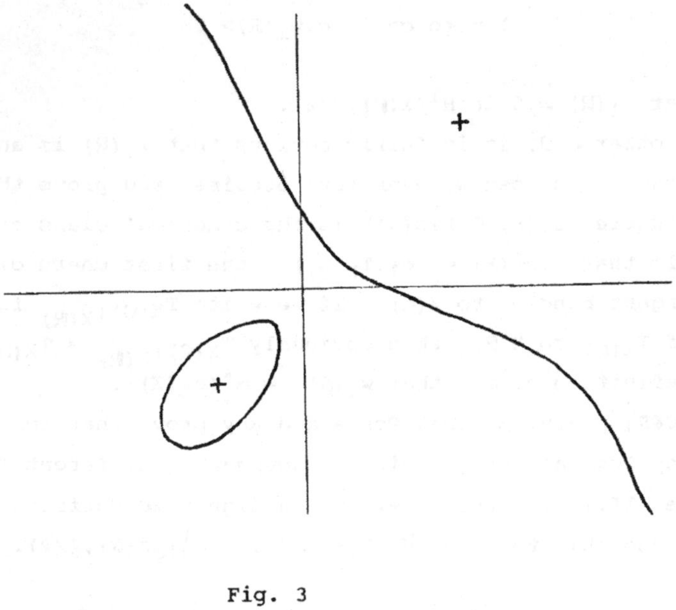

Fig. 3

Let X be the surface in $\mathbb{P}^3$ defined by $p(x,y,z) + w^4z = 0$.

X is smooth and X(ℝ) is a double cover of the region marked "_" in fig. 3, ramified along C(ℝ) and glued at infinity along the line

$z = 0$, $x = -y$. $X(\mathbb{R})$ is non orientable because it is of odd degree in $\mathbb{P}^3$ and it has one handle by construction. Hence it is a Klein bottle.

Now let $\xi = (\xi_1,\ldots,\xi_5,\xi_6,\ldots,\xi_{21}) \in \mathbb{R}^{21}$ be a transcendence basis of degree 21 over $\mathbb{Q}$ and let X_ξ be defined by $(\xi_1x^5 + \xi_2xyz^3 + \xi_3y^5 - \xi_4z^5/32 + \xi_5w^4z) + \xi_6x^4y + \ldots$. Then X_ξ is generic in the sense of Deligne [De$_2$] and we can apply Noether's theorem (see [De$_2$] (1.2.1)) which implies that rank $NS(X_\xi) = 1$.

Using this, we see that $\dim H^2(G,NS(X_\xi)) \leq 1$ (in fact, the considerations of II.§3 show that in this case we must have $\dim H^2(G,NS(X_\xi)) = 1$). On the other hand, because this is an hypersurface in $\mathbb{P}^3$, we have by Lefschetz's theorem $H^1(X_\xi(\mathbb{C}),\mathbb{Z}) = 0$. Applying (1.10) and (2.4) ($\deg X_\xi = 5$) we get :

$$\dim H^1_{alg}(X_\xi(\mathbb{R}),\mathbb{Z}/2) = 1.$$

Now if we choose ξ such that $|1 - \xi_i| \leq \epsilon$ for $1 \leq i \leq 5$, $|\xi_i| \leq \epsilon$ for $6 \leq i \leq 21$. If ϵ is small enough X_ξ will be smooth and $X_\xi(\mathbb{R})$ will have same topological type as $X(\mathbb{R})$. This proves our claim.

In the same line of ideas we can give an example of a surface for which $X(\mathbb{R})$ is not a sphere but $H^1_{alg}(X(\mathbb{R}),\mathbb{Z}/2) = \{0\}$.

The following example is based on an idea of Kucharz (private communication).

Let X be a surface of degree 4 in $\mathbb{P}^3$ such that the real part $X(\mathbb{R})$ is a torus contained in $\mathbb{R}^3$ (this is easy to realise, one can take a surface which is the product of two circles) changing parameters slightly such that the surface becomes smooth and generic as above and the real part stays a torus. We again have, rank $\mathrm{Pic}(X) = 1$. In fact, $\mathrm{Pic}(X)$ is generated by hyperplane sections, that is, by divisors of the form $X \cap H$ where H is a plane in $\mathbb{P}^3$.

But $\alpha'(X \cap H)$ is just $X(\mathbb{R}) \cap H$. Taking H to be the plane at infinity we must have $\alpha^*(X \cap H) = 0$. Since α^* is surjective onto $H^1_{alg}(X(\mathbb{R}),\mathbb{Z}/2)$, this means that $H^1_{alg}(X(\mathbb{R}),\mathbb{Z}/2) = 0$.

This situation has in fact an easy interpretation, the intersection $X(\mathbb{R}) \cap H$ is, when non empty or reduced to points, a plan irreducible quartic curve with two or one components and in the latter case the cycle thus defined is trivial.

We will need, for later use, a lemma of Borel and Haefliger :

(2.5) <u>Lemma</u> : <u>If</u> α <u>is as defined in</u> (1.4) <u>and</u> "(.)" (<u>resp</u>; "< . >") <u>is the intersection form on</u> X (<u>resp. the intersection form</u> mod.2 <u>on</u> $X(\mathbb{R})$) <u>then</u> :

$$(D.D') \equiv \langle\alpha(D).\alpha(D')\rangle \pmod{2}\ .$$

<u>Proof</u> : For a proof in a more general context, see [Bo&Ha]. We give here a proof for surfaces.

We first note that if D' is any divisor in $\mathrm{Div}(X)^G$ (that is, such that $D' = D'^S$), then

$$(D'.D + D^S) = 2(D'.D)$$

by II.(1.1). But in the proof of (1.9) we have seen that $\alpha(D + D^S) = 0$. This shows that (2.5) holds in this case.

To end the proof of (2.5) we only need now to consider the case when both D and D' are irreducible (over $\mathbb{C}$) and defined over $\mathbb{R}$. Also, by definition of the intersection form on X (see for example [Ha] p. 357-360), we can reduce to the case when the intersection of D and D' is transversal. But in this case (2.5) is obvious because the number of complex non-real points of intersection of 2 real curves is always even.

(2.6) <u>Remarks</u> :

(i) If $X(\mathbb{R}) = \emptyset$ we can define α to be the zero map. The above argument (or the original proof given by Borel and Haefliger) still applies in this case and we have, if $X(\mathbb{R}) = \emptyset$, for any pair of S-invariant divisors $(D.D') \equiv 0$ (mod.2). One word of caution is necessary when applying this formula : the divisors must be S-invariant not just of S-invariant class in Pic(X) (I.(4.5) does not hold if $X(\mathbb{R}) = \emptyset$).

(ii) The interest in $H^1_{alg}(X(\mathbb{R}),\mathbb{Z}/2)$ originated with the problem of approximating C^∞-sub-manifolds in real algebraic varieties (see [B&C&R] chap. 11 and 12 for this aspect of the question and a bibliography on the subject). From the point of view of real algebraic geometry it has another interest and as we will try to show in §3 and §4 of this Chapter it plays in real algebraic geometry the role of the Néron-Severi group of classical algebraic geometry.

(iii) We have said above that the inequality $H^1_{alg}(X(\mathbb{R}),\mathbb{Z}/2) \neq H^1(X(\mathbb{R}),\mathbb{Z}/2)$ was generic. In fact the generic situation is $\dim H^1_{alg}(X(\mathbb{R}),\mathbb{Z}/2) \leq 1$. To see this, let $\mathcal{Y}$ be a topological surface (compact, connected and without boundary) in $\mathbb{P}^3(\mathbb{R})$. Since $H^1(\mathbb{P}^3(\mathbb{R}),\mathbb{Z}/2) = H^1_{alg}(\mathbb{P}^3(\mathbb{R}),\mathbb{Z}/2)$, $\mathcal{Y}$ can be approximated by a smooth algebraic surface homeomorphic of $\mathcal{Y}$ (see [B&C&R] 12.4.10 or Benedetti and Tognoli [Be&To]). Making a small change of coefficients if needed we can always choose the surface to be generic (coefficients algebraically independent over $\mathbb{Q}$). If the degree of the surface is more than 3 we apply Noether's theorem and use the same argument as above for our quintic surface.

3. A conjecture of Colliot-Thélène and Sansuc and $H^1_{alg}(X(\mathbb{R}),\mathbb{Z}/2)$ for rational surfaces.

The following definition is established by usage, although it can be confusing :

(3.1) Definition : Let X_o be a surface over $\mathbb{R}$. We will say that X_o is a rational surface if $X = X_o \times_{\mathbb{R}} \mathbb{C}$ is birationally equivalent (over $\mathbb{C}$) to $\mathbb{P}^2_{\mathbb{C}}$. If X is a real surface in the sense of I.(1.15), we will say that X is a rational surface defined over $\mathbb{R}$ if X is birationally equi-

valent (over $\mathbb{C}$*) to* $\mathbb{P}^2_{\mathbb{C}}$. *If* X_o *is birationally equivalent to* $\mathbb{P}^2_{\mathbb{R}}$*, we will say that* X_o *is birationnaly trivial. In this case, we will also say that* $X = X_o \times_{\mathbb{R}} \mathbb{C}$ *is birationally trivial.*

We will need the following consequence of a theorem of Krasnov :

(3.2) *Proposition : Let* X *be a smooth projective real surface. If* X *is a complete intersection (in particular if* X *is a surface in* $\mathbb{P}^3$*) or if* X *is a rational surface and if* $X(\mathbb{R}) \neq \emptyset$*, then :*

$$\dim H^0(X(\mathbb{R}),\mathbb{Z}/2) = (1/2)(\dim H^2(G,H^2(X(\mathbb{C}),\mathbb{Z})) + 2)$$
$$= (1/2)(b_2 - \lambda_2 + 2)$$

$$\dim H^1(X(\mathbb{R}),\mathbb{Z}/2) = \dim H^1(G,H^2(X(\mathbb{C}),\mathbb{Z}))$$
$$= B_2 - b_2 - \lambda_2 \ .$$

Proof : Under the hypotheses of (3.2), the cohomology of $X(\mathbb{C})$ is without torsion and $H^1(X(\mathbb{C}),\mathbb{Z}/2) = 0$. By Krasnov's theorem $A_1.7$, X is a GM-surface. Noting that the hypotheses imply that $b_1 = \lambda_1 = 0$, (3.2) follows from I.(3.13).

We are now in a position to prove :

(3.3) *Theorem* : (Colliot-Thélène and Sansuc conjecture) : *Let* X *be a rational surface defined over* $\mathbb{R}$. *If* $X(\mathbb{R}) \neq \emptyset$ *the number of connected components of* $X(\mathbb{R})$ *is* :

$$^{\#}X(\mathbb{R}) = (1/2)\dim H^1(G,\mathrm{Pic}(X)) + 1 \ .$$

Proof : By (3.2), we have $2^{\#}X(\mathbb{R}) = 2 + \dim H^2(G,H^2(X(\mathbb{C}),\mathbb{Z}))$. On the other hand, since the geometric genus $p_g = h^{0,2}$ of X and the irregularity $q = h^{0,1}$ are zero,

$$H^2(X(\mathbb{C}),\mathbb{Z}) \cong NS(X) \cong \mathrm{Pic}(X)$$

(see for example [Gr&Ha] p. 163).

We note that the above are not isomorphisms of G-modules but that $H^2(X(\mathbb{C}),\mathbb{Z})(1) \cong \mathrm{Pic}(X)$ is (see I.§4). By I.(3.7) this implies
$\dim H^2(G,H^2(X,(\mathbb{C}),\mathbb{Z})) = \dim H^1(G,H^2(X(\mathbb{C}),\mathbb{Z})(1)) = \dim H^1(G,\mathrm{Pic}(X))$

which proves the theorem by (3.2).

Our next result is :

(3.4) <u>Theorem</u> : <u>If</u> X <u>is a rational surface defined over</u> $\mathbb{R}$, <u>we have</u> :

$$H^1_{alg}(X(\mathbb{R}),\mathbb{Z}/2) = H^1(X(\mathbb{R}),\mathbb{Z}/2) .$$

<u>Proof</u> : We first note that by (3.2) we have :

$$\dim H^1(X(\mathbb{R}),\mathbb{Z}/2) = \dim H^1(G,H^2(X(\mathbb{C}),\mathbb{Z})).$$

For the same reasons as in the proof of (3.3), this implies that :

$$\dim H^1(X(\mathbb{C}),\mathbb{Z}/2) = \dim H^2(G,\mathrm{Pic}(X)).$$

We have going to prove that, under the hypotheses of (3.4), we also have $\dim H^1_{alg}(X(\mathbb{R}),\mathbb{Z}/2) = \dim H^2(G,\mathrm{Pic}(X))$.

For this we choose a basis $\{d_i\}$ of Pic(X) of the form defined in I.(3.5). We note that by definition, if $r = \mathrm{rank}\,\mathrm{Pic}(X)^G$, then the r first elements of this basis form a basis of $\mathrm{Pic}(X)^G = \mathrm{Ker}(1-S)$ and that if λ is the Comessatti characteristic for the action of S on Pic(X), then :

(3.5)
$$d_i \in (1+S)\mathrm{Pic}(X) \quad \text{if } 1 \leqslant i \leqslant \lambda$$
$$d_j \notin (1+S)\mathrm{Pic}(X) \quad \text{if } \lambda < j \leqslant r .$$

Recalling that for x and y in Pic(X),

$$(x.y) = (S(x).S(y)) \qquad \text{(see II.(1.1))}$$

(this makes sense since the intersection form only depends on the linear equivalence class), we have for all i, $1 \leqslant i \leqslant \lambda$, d_i as above and all $d \in \mathrm{Pic}(X)^G$:

$$(3.6) \qquad (d_i.d) = (d'.d) + (d'^S.d) = 2(d'.d) \equiv 0 \pmod{2} .$$

The hypothesis of (3.4) imply that $\mathrm{Pic}(X) \cong H^2(X(\mathbb{C}),\mathbb{Z})(1)$ (see proof of (3.3)). This in turn implies :

(i) the λ appearing in (3.5) is the same as λ_2 ;
(ii) the cup product form Q on $H^2(X(\mathbb{C}),\mathbb{Z})$ is identified with the intersection form on Pic(X).

(i) and (ii) combined with II.(1.2) imply that the determinant of the intersection form restricted to $\mathrm{Pic}(X)^G$ ($\cong H^2(X(\mathbb{C}),\mathbb{Z})(1)^G$) is equal to 2^λ.

Expressing the matrix of the intersection form restricted to $\mathrm{Pic}(X)^G$ in terms of the basis $\{d_i\}_{1\leqslant i\leqslant r}$ mentioned above, we see that the evaluation of the determinant together with (3.6) imply that for all j, $\lambda < j \leqslant r$, there exists a k such that

$$(d_j.d_k) \equiv 1 \pmod{2} .$$

If we apply (2.5) to this we get :

$$<\alpha^*(d_j).\alpha^*(d_k)> = 1$$

This implies that for all j, $\lambda < j \leqslant r$: $\alpha^*(d_j) \neq 0$.

If we replace d_{j_1} by $d_{j_1}+\ldots+d_{j_s}$ (the j_i's distinct and $\lambda < j_i \leqslant r$) we obtain again a basis of $\mathrm{Pic}(X)^G$ satisfying (3.5). For the same reasons, $\alpha^*(d_{j_1}+\ldots+d_{j_s}) \neq 0$. In other words the $\alpha^*(d_j)$, $\lambda < j \leqslant r$, are linearly independant in $H^1_{alg}(X(\mathbb{R}),\mathbb{Z}/2)$.

But now we are done, because this implies that

$$\dim H^1_{alg}(X(\mathbb{R}),\mathbb{Z}/2) \geqslant r - \lambda = \dim H^2(G,\mathrm{Pic}(X))$$

and hence

$$\dim H^1_{alg}(X(\mathbb{R}),\mathbb{Z}/2) = \dim H^2(G,\mathrm{Pic}(X))$$

by (1.10), as announced.

4. The topology of $X(\mathbb{R})$ for a rational surface I.

We assume throughout this § that X is a smooth rational surface. This allows for the identifications $\mathrm{Pic}(X) = H^2(X(\mathbb{C}),\mathbb{Z})(1)$ (see §3) that we will also use throughout.

We consider once more the cup-product form Q on $H^2(X(\mathbb{C}),\mathbb{Z})$ and we recall that $Q(S(x),S(y)) = Q(x,y)$.

Let $x = z-S(z) \in (1-S)H^2(X(\mathbb{C}),\mathbb{Z}) \subset H^2(X(\mathbb{C}),\mathbb{Z})(1)^G$; then for all $y \in H^2(X(\mathbb{C}),\mathbb{Z})(1)^G$ (hence such that $S(y) = -y$) we have :

$$Q(x,y) = Q(z,y) - Q(S(z),y) = 2Q(z,y) \equiv 0 \pmod{2} .$$

This implies that Q induces a bilinear form with values in $\mathbb{Z}/2$ on :

$$H^2(X(\mathbb{C}),\mathbb{Z})(1)^G/(1-S)H^2(X(\mathbb{C}),\mathbb{Z}) = H^2(G,H^2(X(\mathbb{C}),\mathbb{Z})(1)).$$

We denote Q_2 this form.

(4.1) Proposition : If X is a rational surface defined over $\mathbb{R}$ and if $X(\mathbb{R}) \neq \emptyset$, then the morphism β defined in (1.8) is an isometry between $(H^2(G,H^2(X(\mathbb{C}),\mathbb{Z})(1)),Q_2)$ and $(H^1(X(\mathbb{R}),\mathbb{Z}/2),<.>)$.

<u>Proof</u> : We first recall that $H^2(G,H^2(X(\mathbb{C}),\mathbb{Z})(1)) \cong H^1(G,H^2(X(\mathbb{C}),\mathbb{Z}))$ by definition (I.(3.6.1) and I.(4.8)) and $H^1(G,H^2(X(\mathbb{C}),\mathbb{Z})) \cong H^1(X(\mathbb{R}),\mathbb{Z}/2)$ by (3.2).

Since $\beta : H^2(G,\mathrm{Pic}(X)) = H^2(G,H^2(X(\mathbb{C}),\mathbb{Z})(1)) \longrightarrow H^1_{alg}(X(\mathbb{R}),\mathbb{Z}/2)$ is surjective by construction and $H^1_{alg}(X(\mathbb{R}),\mathbb{Z}/2) = H^1(X(\mathbb{R}),\mathbb{Z}/2)$ by (3.4), we conclude that β is an isomorphism.

Let d and d' be in $\mathrm{Pic}(X)$ and $\tilde{d}$ and $\tilde{d}'$ their classes in $H^2(G,\mathrm{Pic}(X))$. We have $Q(d,d') \equiv Q_2(\tilde{d},\tilde{d}')$ (mod.2) by definition.

On the other hand, by (2.5) and by definition of β :

$$Q(d,d') \equiv \langle\alpha^*(d).\alpha^*(d')\rangle = \langle\beta(\tilde{d}).\beta(\tilde{d}')\rangle \quad (\mathrm{mod.}2) .$$

This proves that β is an isometry.

(4.1.1) <u>Remark</u> : If we write $(.)$ for the intersection form in $\mathrm{Pic}(X)$, and $(.)_2$ for the form induced in $H^2(G,\mathrm{Pic}(X))$ we can reformulate (4.1) by saying that there is an isometry between $(H^2(G,\mathrm{Pic}(X)),(.)_2)$ and $(H^1(X(\mathbb{R}),\mathbb{Z}/2),\langle.\rangle)$.

(4.2) <u>Proposition</u> : <u>Let</u> X <u>be a smooth projective real surface</u>. <u>We have the following implication</u>.

(i) <u>There exists an</u> S-<u>invariant divisor such that</u> $D^2 \equiv 1$ (mod.2) <u>implies</u> :

(ii) $X(\mathbb{R}) \neq \emptyset$ <u>and</u> $X(\mathbb{R})$ <u>has a non orientable component</u>.

<u>If moreover</u> X <u>is a rational surface and</u> $X(\mathbb{R}) \neq \emptyset$, <u>then</u> (i) <u>and</u> (ii) <u>are equivalent to</u> :

(iii) <u>there exists</u> $c \in H^2(X(\mathbb{C}),\mathbb{Z})(1)^G = \mathrm{Ker}(1+S)$ <u>such that</u> $Q(c,c) \equiv 1$ (mod. 2).

<u>Proof</u> : We first recall that if $X(\mathbb{R}) = \emptyset$ then for all S-invariant divisors we have $D^2 \equiv 0$ (mod.2) (see (2.6)(i)).

Second, we recall that by lemma (1.1) of [Mi&Hu] p. 101, $X(\mathbb{R})$ has a non orientable component if and only if there exists $\gamma \in H^1(X(\mathbb{R}),\mathbb{Z}/2)$ such that $\langle\gamma,\gamma\rangle = 1$. With this in mind, the implication (i) $\Longrightarrow$ (ii) follows in the general case from (2.5) and (2.6).

If X is rational, then, by (3.4), we can find for every $c \in H^1(X(\mathbb{R}),\mathbb{Z}/2)$ a divisor D, S-invariant, such that the class of $\alpha(D)$ is c. (2.5) and the above characterization of non orientable surfaces prove the implication (ii) $\Longrightarrow$ (i) in this case.

Finally, if X is rational then $\mathrm{Pic}(X)^G = H^2(X(\mathbb{C}),\mathbb{Z})(1)^G$ and (i) $\Longleftrightarrow$ (iii) follows from the facts that if $X(\mathbb{R}) \neq \emptyset$ the map $\mathrm{Div}(X)^G \longrightarrow \mathrm{Pic}(X)^G$ is surjective (see I.(4.5)) and that the intersection form only depends on the linear class.

For orientable components we have :

(4.3) <u>Proposition</u> : <u>If</u> X <u>is a rational surface defined over</u> $\mathbb{R}$ <u>and if</u> X_1 <u>is an orientable component of</u> $X(\mathbb{R})$, <u>then</u> X_1 <u>is homeomorphic to the sphere</u> S^2 <u>or to the torus</u> T_1. <u>In the latter case</u>, $X(\mathbb{R})$ <u>is connected and we have</u> $X(\mathbb{R}) = T_1$.

<u>Proof</u> : We first recall that the signature of Q^+ (the restriction of the cup-product form to the invariant part under S of $H^2(X(\mathbb{C}),\mathbb{Z})$ is by II.(1.2) $(h^{0,2},b_2-h^{0,2})$ hence here, since X is rational, $(0,b_2)$. In other words, Q^+ is negative definite. Now let x_1 be the fundamental class of X_1 in $H^2(X(\mathbb{C}),\mathbb{Z})$. x_1 is obviously in $H^2(X(\mathbb{C}),\mathbb{Z})^G$. By our first remark this implies that we must either have $x_1 = 0$ or $Q(x_1,x_1) < 0$.

Since X_1 is orientable $Q(x_1,x_1)$ is equal to the self-intersection of X_1 in $X(\mathbb{C})$. But by definition (see Hirsh [Hi] p. 132) the self-intersection of X_1 in $X(\mathbb{C})$ is equal to the self- intersection of X_1 in the normal bundle $N_{X(\mathbb{C})|X_1}$ of X_1 in $X(\mathbb{C})$ (see also [Mi&St] p. 119).

Now let $x \in X_1 \subset X(\mathbb{R})$ and let (u_1,u_2) be a basis of $T_{X(\mathbb{R}),x}$. Since $x \in X(\mathbb{R})$, $T_{X(\mathbb{C}),x} = T_{X(\mathbb{R}),x} \otimes_{\mathbb{R}} \mathbb{C}$ and (u_1,iu_1,u_2,iu_2) is a basis of $T_{X(\mathbb{C}),x}$. In this context, (iu_1,iu_2) is a basis of $N_{X(\mathbb{C})|X(\mathbb{R}),x}$ (the normal space). Actually, since the natural orientation of $X(\mathbb{C})$ is given by (u_1, iu_1, u_2, iu_2) an oriented basis of $N_{X(\mathbb{C})|X(\mathbb{R}),x}$ is given by (iu_2, iu_1).

This means that multiplication by $i = \sqrt{-1}$ in $T_{X(\mathbb{C})}$ induces an orientation reversing isomorphism between $T_{X(\mathbb{R})}$ and $N_{X(\mathbb{C})|X(\mathbb{R})}$. The same is of course true for X_1 and this implies that $Q(x_1, x_1)$ is equal to minus the self-intersection of X_1 in its tangent bundle. But this last self-intersection is nothing but the Euler characteristic $\chi(X_1)$ of X_1 (see Hirsh [Hi] p. 13). By the privious computation of Q this means that $\chi(X_1) \geq 0$ and proves the first assertion of (4.3).

To prove the last assertion we use an argument of Risler [Ri_2] p 161. We consider once more the Smith sequence and use again the notations of I.§3, $\pi : X(\mathbb{C}) \longrightarrow Y = X(\mathbb{C})/G$ and $\mathcal{F} = \pi_*\mathbb{Z}/2$. By the description of $\mathcal{F}$ given in I.§3 (see also proof of II.(5.4)), we see that $H^i(Y, (1+S)\mathcal{F}) = H^i_c(Y\setminus X(\mathbb{R}), \mathbb{Z}/2)$ (cohomology with compact support).

Since X is rational, $H^1(X(\mathbb{C}), \mathbb{Z}/2) = H^3(X(\mathbb{C}), \mathbb{Z}/2) = 0$. By the Smith sequence (I.(3.1)), we therefore have :

$$0 \longrightarrow H^3_c(Y\setminus X(\mathbb{R}), \mathbb{Z}/2) \longrightarrow H^4_c(Y\setminus X(\mathbb{R}), \mathbb{Z}/2) \longrightarrow$$
$$H^4(X(\mathbb{C}), \mathbb{Z}/2) \longrightarrow H^4_c(Y\setminus X(\mathbb{R}), \mathbb{Z}/2) \longrightarrow 0$$

From this we see that $H^4(X(\mathbb{C}), \mathbb{Z}/2) = \mathbb{Z}/2$ implies that $H^4_c(Y\setminus X(\mathbb{R}), \mathbb{Z}/2) = \mathbb{Z}/2$ and hence also $H^3_c(Y\setminus X(\mathbb{R}), \mathbb{Z}/2) = \mathbb{Z}/2$.

Consider now the exact sequence of cohomology with compact support and coefficients in $\mathbb{Z}/2$ associated to the inclusion $i = X(\mathbb{R}) \hookrightarrow Y$. The part we are interested in is :

$$(4.4) \qquad H^2(Y, \mathbb{Z}/2) \xrightarrow{\ i^*\ } H^2(X(\mathbb{R}), \mathbb{Z}/2) \longrightarrow H^3_c(Y\setminus X(\mathbb{R}), \mathbb{Z}/2) .$$

Let $\{X_k\}_k$ be the components of $X(\mathbb{R})$ and consider the map :

(4.5) $$i_* : H_2(X(\mathbb{R}),\mathbb{Z}/2) \longrightarrow H_2(Y,\mathbb{Z}/2) .$$

i_* maps the fundamental homology class of X_k in $H_2(X(\mathbb{R}),\mathbb{Z}/2)$ onto the fundamental homology class in $H_2(Y,\mathbb{Z}/2)$. This map is dual to the map i^* of (4.4), so by the above computation we have :

(4.6) $$\dim \operatorname{Ker} i_* = \dim \operatorname{Coker} i^* \leqslant \dim H_c^3(Y\setminus X(\mathbb{R}),\mathbb{Z}/2) = 1 .$$

To end the proof of (4.3), we note that on the one hand $\pi : X(\mathbb{C}) \longrightarrow Y$ is a double cover ramified along $X(\mathbb{R})$. This implies that if $X(\mathbb{R})$ is orientable, the fundamental class of $X(\mathbb{R})$ in Y is 2-divisible and that, in the general case the fundamental class mod.2 of $X(\mathbb{R})$ in Y is zero (see Hirzebruch [Hz] p. 259-260). In any case, the fundamental class of $X(\mathbb{R})$ in $H^2(Y,\mathbb{Z}/2)$ is zero and by duality also in $H_2(Y,\mathbb{Z}/2)$.

On the other hand, if $X(\mathbb{R})$ has a component X_1 homeomorphic to a torus T_1 then we have seen in the first part of the proof of (4.3) that the fundamental class of X_1 in $H^2(X(\mathbb{C}),\mathbb{Z})$ is zero. This implies that it is also zero in $H^2(Y,\mathbb{Z})$ and hence also in $H^2(Y,\mathbb{Z}/2)$.

This proves (4.3) because if $X_1 \neq X(\mathbb{R})$ then the fundamental classes of X_1 and $X(\mathbb{R})$ in $H_2(X(\mathbb{R}),\mathbb{Z}/2)$ are linearly independant, but by (4.6) they cannot be both maped to zero in $H_2(Y,\mathbb{Z}/2)$. Hence if $X_1 \neq X(\mathbb{R})$, $\chi(X_1) > 0$.

(4.7) <u>Remarks</u> :

(i) In constrast to (4.3) we note that using monoidal transformations one can find as components of the real part of a rational surface any type of non-orientable topological surface. Moreover we will see as a consequence of Chapter VI. that we can in fact find for the real part of a rational surface any combination of non-orientable components (see VI.(6.2)).

(ii) In §3 and §4 we have not fully used the fact that the surfaces were rational. We have only used the fact that $H^*(X(\mathbb{C}),\mathbb{Z})$ was without 2-torsion and that $p_g = q = 0$ (that is, $h^{0,2} = h^{0,1} = 0$).

This implies that all the results of §3 and §4 still hold for surfaces with these properties. This is the case for certain surfaces of general type (see for example [B&P&V] p. 234-237).

Bibliographical Notes :

For general references on algebraic cycles see [B&C&R].

(2.4) can be found in [B&K&S].

The conjecture of Colliot-Thélène and Sansuc (3.3) can be found in [Co&Sa] p. 970.

The proof of (4.3) is inspired of Risler [Ri_2].

For the rest, most of material in this chapter, except the examples given in §2, comes from [Si_5].

IV. DIGRESSION ON REAL ABELIAN VARIETIES AND CLASSIFICATION OF REAL ABELIAN SURFACES

1. Real structure on the Albanese and on the Picard variety

Let X be smooth projective real variety with real structure (X,S). We consider the exact sequence defining the Albanese of X (or rather of $X(\mathbb{C})$) ;

$$(1.1) \quad 0 \longrightarrow H_1(X(\mathbb{C}),\mathbb{Z})_f \longrightarrow H^0(X(\mathbb{C}),\Omega^1_{X(\mathbb{C})})^* \longrightarrow \mathrm{Alb}(X) \longrightarrow 0 ,$$

where $H_1(X(\mathbb{C}),\mathbb{Z})_f$ is $H_1(X(\mathbb{C}),\mathbb{Z})$ modulo torsion, $\Omega^1_{X(\mathbb{C})}$ the sheaf of holomorphic differential 1-forms, " * " means the dual and the first map is

$$(1.2) \qquad \gamma \longrightarrow [\omega \longrightarrow \textstyle\int_\gamma \omega] .$$

The action of S on $X(\mathbb{C})$ induces on the first two groups of the sequence (1.1) a natural G-module structure for which (1.2) defines a G-map. In other words, (1.1) endows $\mathrm{Alb}(X)$ with a natural G-structure for which (1.1) becomes an exact sequence of G-modules.

G operates on $\Omega^1_{X(\mathbb{C})}$ is a manner very similar to the way it operates on $\mathcal{O}_{X(\mathbb{C})}$. More precisely, if we choose local coordinates satisfying I.(1.11) and if $\omega = \Sigma f_i dz_i$ is a holomorphic 1-form, then $\omega^S = \Sigma\, j \circ f_i \circ S\, dz_i$ (with j complex conjugation in $\mathbb{C}$). It is easy to see that for all 1-forms ω, $\omega_1 = 1/2(\omega+\omega^S)$ and $\omega_2 = (\sqrt{-1}/2)(\omega-\omega^S)$ are in $H^0(X(\mathbb{C}),\Omega^1_{X(\mathbb{C})})^G$.

Using this we note that, if the irregularity of X is q $(= \dim_{\mathbb{C}} H^0(X(\mathbb{C}),\Omega^1_X))$, we have :

$$(1.5) \qquad H^0(X(\mathbb{C}),\Omega^1_{X(\mathbb{C})})^G = \mathbb{R}^q$$

and hence also : $H^0(X(\mathbb{C}),\Omega^1_{X(\mathbb{C})})^{*G} = \mathbb{R}^q$

The same argument also proves that :

(1.6) $$H^i(G,H^0(X(\mathbb{C}),\Omega^1_{X(\mathbb{C})})^*) = 0 \quad \text{for all } i > 0.$$

By (1.6), the exact sequence of Galois cohomology associated to (1.1) is :

(1.7) $$0 \longrightarrow H_1(X(\mathbb{C}),\mathbb{Z})^G_f \longrightarrow H^0(X(\mathbb{C}),\Omega^1_{X(\mathbb{C})})^{*G} \longrightarrow \mathrm{Alb}(X)^G \longrightarrow H^1(G,H_1(X(\mathbb{C}),\mathbb{Z})_f) \longrightarrow 0$$

By the isomorphism $H_1(X(\mathbb{C}),\mathbb{Z})_f = H^1(X(\mathbb{C}),\mathbb{Z})^*$ and I.(3.7) we have :

(1.8) $$H_1(X(\mathbb{C}),\mathbb{Z})^G_f = \mathbb{Z}^q$$

and $$H^1(G,H_1(X(\mathbb{C}),\mathbb{Z})_f) = (\mathbb{Z}/2)^{q-\lambda_1} ,$$

λ_1 as in I.(3.12).

(1.5) and (1.8) applied to (1.7) yield :

(1.9) <u>Proposition</u> : <u>Let</u> (X,S) <u>be a smooth projective real variety</u>. <u>The Albanese variety</u> Alb(X) <u>has a natural real structure and</u> $\mathrm{Alb}(X)^G = \mathrm{Alb}(X)(\mathbb{R})$ <u>is a compact real Lie group</u>. <u>As such</u> :

$$\mathrm{Alb}(X)(\mathbb{R}) = (\mathbb{R}/\mathbb{Z})^q \times (\mathbb{Z}/2)^{q-\lambda_1}$$

<u>where</u> q <u>is the irregularity of</u> X <u>and</u> $\lambda_1 = \mathrm{rank}(1+S)H_1(X(\mathbb{C}),\mathbb{Z})_f$.

For the Picard variety we have already noted in III.§1 the exact sequence of G-modules :

$$0 \longrightarrow H^1(X(\mathbb{C}),\mathbb{Z})(1) \longrightarrow H^1(X(\mathbb{C}),\mathcal{O}_{X(\mathbb{C})}) \longrightarrow \mathrm{Pic}^0(X) \longrightarrow 0 .$$

The corresponding sequence of Galois cohomology is :

$$(1.10)\quad 0 \longrightarrow H^1(X(\mathbb{C}),\mathbb{Z})(1)^G \longrightarrow H^1(X(\mathbb{C}),\mathcal{O}_{X(\mathbb{C})})^G \longrightarrow \mathrm{Pic}^o(X)^G \longrightarrow H^1(G,H^1(X(\mathbb{C}),\mathbb{Z})(1)) \longrightarrow 0$$

Recalling that $b_1 = B_1/2$ (see I.(2.5)) we have by the same computation as for (1.9) :

(1.11) Proposition : *If* (X,S) *is a smooth projective real variety, then the Picard variety* $\mathrm{Pic}^o(X)$ *has a natural real structure and* :

$$\mathrm{Pic}^o(X)^G = \mathrm{Pic}^o(X)(\mathbb{R}) = (\mathbb{R}/\mathbb{Z})^q \times (\mathbb{Z}/2)^{q-\lambda_1}$$

where q *and* λ_1 *are as in* (1.9) (*or equivalently* λ_1 *is an in* I.(3.12)).

(1.12) Remark :

We recall that if X is an abelian variety then X = Alb(X). (1.9) implies in this case that if X is an abelian variety then the topology of $X(\mathbb{R})$ is completely determined by the Galois structure of $H_1(X(\mathbb{C}),\mathbb{Z})_f$ or equivalently by that of $H^1(X(\mathbb{C}),\mathbb{Z})$. A fortiori it is determined by the Galois structure of $H^*(X(\mathbb{C}),\mathbb{Z})$. This suggests that a real abelian variety is a GM-variety (see I.(3.11)). This is in fact the case and can be easily proved using I.(3.5) and (1.9) (see Krasnov [Kr] for a proof).

2. The period matrix associated to a real structure and the Albanese map.

In all of this § X will be a smooth projective real variety with real structure (X,S). Also to simplify notations we will write λ in place of λ_1 (λ_1 as in (1.9) above).

By I.(3.5) and I.(2.5) we can choose a basis

$$(2.1)\qquad (a_1,\ldots,a_q,b_1,\ldots,b_\lambda,c_{\lambda+1},\ldots,c_q)$$

of $H_1(X(\mathbb{C}),\mathbb{Z})_f$ such that : $S(a_i) = a_i$, $S(b_j) = a_j-b_j$ and $S(c_k) = -c_k$. We will say that such a basis is semi-real (or S-semi real).

On the other hand, by (1.5), a basis of $H^0(X(\mathbb{C}),\Omega^1_{X(\mathbb{C})})^G$ over $\mathbb{R}$ is a basis of $H^0(X(\mathbb{C}),\Omega^1_{X(\mathbb{C})})$ over $\mathbb{C}$. We will call such a basis real.

We fix a semi-real basis (a_i,b_j,c_k) of $H_1(X(\mathbb{C}),\mathbb{Z})_f$ and a real basis (ω_i) of $H^0(X(\mathbb{C}),\Omega^1_{X(\mathbb{C})})$.

If γ is a 1-cycle on $X(\mathbb{C})$ and ω a 1-form we have :

$$\text{(2.2)} \qquad \int_{S(\gamma)} \omega^S = \overline{\int_\gamma \omega} \quad ,$$

where $\overline{}$ is complex conjugation. In particular, if the 1-form ω is in $H^0(X(\mathbb{C}),\Omega^1_{X(\mathbb{C})})^G$ and a_i is as above, then

$$\int_{a_i} \omega \in \mathbb{R} \ .$$

Now since the $\left\{\int_{a_i} \omega_j\right\}_{1 \leq j \leq q}$, as vectors in $\mathbb{C}^q$, are linearly independent (and hence also as vectors in $\mathbb{R}^q$) we can normalize the real basis (ω_i) in such a way that :

$$\int_{a_i} \omega_i = 1 \quad , \quad \int_{a_i} \omega_j = 0 \quad \text{if } i \neq j \ .$$

If we do this (2.1) and (2.2) imply

$$\int_{b_j} \omega_j = 1/2 + \sqrt{-1}\, t_{jj} \qquad ,$$

$$\int_{b_j} \omega_i = \sqrt{-1}\, t_{ij} \qquad \text{for } i \neq j \ ,$$

$$\int_{c_k} \omega_h = \sqrt{-1}\, t_{hk}$$

where the $t_{ij} \in \mathbb{R}$.

(2.3) <u>Theorem</u> (Comessatti) : <u>Let</u> X <u>be a complex abelian variety</u>. <u>There</u>

exists a real structure (X,S) *on* X *if and only if* X *has a period matrix of the form* :

$$\Omega = (I_q, 1/2 \Lambda + iT),$$

where I_q *is the identity matrix*, $T \in GL_q(\mathbb{R})$, $i^2 = -1$ *and* Λ *is of the form* $\begin{pmatrix} I_\lambda & 0 \\ 0 & 0 \end{pmatrix}$.

If (X,S) *has real points* (X,S) *is real equivalent to* $(\mathbb{C}^q/[\Omega],\sigma)$, *where* σ *is the involution induced by complex conjugation in* $\mathbb{C}^q$, Ω *is a period matrix for* X *of the above form and* $[\Omega]$ *is the lattice in* $\mathbb{C}^q$ *generated by the columns of* Ω.

If the real part of (X,S) *is empty*, (X,S) *is real equivalent to* $(\mathbb{C}^q/[\Omega],t_a\sigma)$, *where* t_a *is translation by an element* a.

Proof : Recalling that a period matrix is by definition a matrix $\left(\int_{a_i} \omega_j\right)_{i,j}$ with $\{a_i\}$ a basis of $H_1(X(\mathbb{C}),\mathbb{Z})_j$ and $\{\omega_j\}$ a basis of $H^0(X(\mathbb{C}),\Omega^1_{X(\mathbb{C})})$, we have already proved the "only if" part of the first statement of the theorem. To prove the "if" part we only need to note that if the period matrix is of the indicated form, then the lattice $[\Omega]$ is globally invariant under complex conjugation. In other words complex conjugation induces an anti-holomorphic involution on $\mathbb{C}^q/[\Omega]$ which is isomorphic over $\mathbb{C}$ to $X(\mathbb{C})$.

To prove the last two statements we will need :

(2.4) *Lemma* : *Let* (X,S) *be a real, smooth and projective, variety. If* $X(\mathbb{R}) \neq \emptyset$, *then for any* $x \in X(\mathbb{R})$ *the Albanese map* $f_x : X \longrightarrow Alb(X)$ (Alb(X) *endowed with its natural real structure*) *is defined over* $\mathbb{R}$.

If $(ALb(X),\tilde{S})$ *is the natural real structure on* Alb(X) *and if* x *is any point in* $X(\mathbb{C})$, *then* $a = f_x(S(x))$ *is pure imaginary for* $\tilde{S}$ (*that is*, $\tilde{S}(a) = -a$) *and* $f_x(S(y)) = \tilde{S}f_x(y) + a$.

Proof : Clearly (2.2) holds if γ is any path in $X(\mathbb{C})$. Since

$$f_x(y) = \left(\left(\int_x^y \omega_i\right)_{1\leqslant i\leqslant q} \text{mod. periods}\right),$$

we have, choosing a real basis for the holomorphic 1-forms :

$$f_{S(x)}(S(y)) = \left(\left(\int_{S(x)}^{S(y)} \omega_i\right)_i \text{ mod. periods}\right)$$

$$= \left(\left(\overline{\int_x^y \omega_i}\right)_i \text{ mod. periods}\right) = \tilde{S}(f_x(y)) \ .$$

From this we see that if $x \in X(\mathbb{R})$, then $f_x(S(y)) = \tilde{S}(f_x(y))$ and f_x is real for the real structures involved. On the other hand, if x is any point in $X(\mathbb{C})$

$$\tilde{S}(f_x(S(x))) = f_{S(x)}(x) = -f_x(S(x))$$

This proves that $f_x(S(x))$ is pure imaginary. The obvious

$$\int_x^{S(y)} \omega_i = \int_x^{S(x)} \omega_i + \int_{S(x)}^{S(y)} \omega_i \text{ (mod. periods)}$$

ends the proof of (2.4).

__End of proof of__ (2.3) : If X is an abelian variety, then for any $x \in X(\mathbb{C})$ the Albanese map f_x is an isomorphism. If $x \in X(\mathbb{R})$, (2.4) proves that this isomorphism is real (or otherwise said, that f_x satisfies condition I.(1.5)). On the other hand, if x is any point in $X(\mathbb{C})$, (2.4) proves that $f_x(S(x)) = a$ is pure imaginary for $\tilde{S}$. This implies that $t_a\tilde{S}(z) = \tilde{S}(z) + a$ is an involution and hence defines a real structure. By construction $t_a\tilde{S} = t_a\sigma$ is equivalent to S and this ends the proof of (2.3).

3. __Real polarization on a real abelian variety__.

We will call polarization on an abelian variety X, the first Chern class of an ample divisor on X. Using of the classical identifications,

$$X(\mathbb{C}) = \mathbb{C}^n/H_1(X(\mathbb{C}),\mathbb{Z})$$

$$H^2(X(\mathbb{C}),\mathbb{Z}) = \Lambda^2 H^1(X(\mathbb{C}),\mathbb{Z})$$

$$= \{\text{group of alternating 2-forms on } H_1(X(\mathbb{C}),\mathbb{Z})\}$$

(see for example Mumford [Mu_2] p. 3), it is equivalent to define a polarization to be an alternating 2-form E on $H_1(X(\mathbb{C}),\mathbb{Z})$ such that :
- the $\mathbb{R}$-linear extension of E to $\mathbb{C}^n$ satisfies :

(3.1) $$E(ix,iy) = E(x,y) \ ;$$

- The hermitian form ($\mathbb{C}$-linear in the second variable) :

(3.2) $$H(x,y) = E(x,iy) + iE(x,y)$$

is positive definite on $\mathbb{C}^n$ (cf. [Mu_2] p. 18-19).

Let (X,S) be a real structure on X. We will say that a polarization E is defined over $\mathbb{R}$ or S-real if :

(3.3) $$E(S_*(x),S_*(y)) = -E(x,y) \ ,$$

where E is identified with an alternating 2-form on $H_1(X(\mathbb{C}),\mathbb{Z})$.

This definition is motivated by :

(3.4) Theorem : Let (X,S) be a real abelian variety with $X(\mathbb{R}) \neq \emptyset$. Let E be the Chern class of a divisor, considered as an alternating 2-form on $H_1(X(\mathbb{C}),\mathbb{Z})$. Then E is the Chern class of an S-invariant (or S-real) divisor if and only if E satisfies :

$$E(S_*(x),S_*(y)) = -E(x,y) \ .$$

Proof : The fact that the condition is necessary follows from

I.(4.12).

To show that the condition is sufficient, first note that since E is the Chern class of a divisor it satisfies condition (3.1) and that as a consequence the form H defined by (3.2) is hermitian (see [Mu_2]). Note also that since $X(\mathbb{R}) \neq \emptyset$, it is sufficient by I.(4.5) to show that E is the Chern class of an S-invariant point of Pic(X) or in other words that there exists an S-invariant line bundle L such that $c_1(L) = E$.

To construct such an L, we are going to use the classical description of line bundles on an abelian variety (see [Mu_2] p. 13-22).

Let $X(\mathbb{C}) = V/\Gamma$ with $V = \mathbb{C}^n$ and Γ a lattice in V. Let $\alpha : \Gamma \to \mathbb{C}^*$ satisfy :

$$(3.5) \qquad \alpha(u+u') = e^{i\pi E(u,u')}.\alpha(u).\alpha(u') \qquad \text{for } u \text{ and } u' \text{ in } \Gamma.$$

Let :

$$e_u(z) = \alpha(u)\ e^{\pi(H(u,z)+\frac{1}{2}H(u,u))}$$

(H defined by (3.2)). Then the line bundle :

$$L = \mathbb{C} \times V/\sim ,$$

where :

$$(\xi,z) \sim (e_u(z).\xi, z+u) \qquad \text{for } u \in \Gamma,$$

has Chern class E.

To find such an L real, note that by theorem (2.3) one can always choose Γ such that S is induced by complex conjugation in $\mathbb{C}^n$ (this implies in particular that Γ is globally invariant under the action of complex conjugation).

Under this hypothesis, L real corresponds to :

$$(\bar{\xi},\bar{z}) \sim (\overline{e_u(z).\xi},\overline{z+u}) = (\overline{e_u(z)}.\bar{\xi},\ \bar{z}+\bar{u}) ,$$

or

$$e_{\bar{u}}(\bar{z}) = \overline{e_u(z)} .$$

S being induced by complex conjugation, the condition on E becomes :

$$E(\bar{u},\bar{v}) = -E(u,v) .$$

For H we then have : $H(\bar{u},\bar{v}) = E(\bar{u},i\bar{v}) + iE(\bar{u},\bar{v})$
$= -E(u,-iv) - iE(u,v)$
$= E(u,iv) - iE(u,v) = \overline{H(u,v)}$,

and hence :

$$e^{\pi(H(\bar{u},\bar{v})+\frac{1}{2}H(\bar{u},\bar{u}))} = \overline{e^{\pi(H(u,v)+\frac{1}{2}H(u,u))}} .$$

This means that to prove (3.4) we only need to find an $\alpha : \Gamma \to \mathbb{C}^*$ satisfying (3.5) and such that $\alpha(\bar{u}) = \overline{\alpha(u)}$.

Since, by (3.5), α is entirely defined by its values on a basis $\{u_i\}$ of the $\mathbb{Z}$-module Γ and since by the conditions on E we have, if $\alpha(\bar{u}) = \overline{\alpha(u)}$ and $\alpha(\bar{v}) = \overline{\alpha(v)}$:

$$\alpha(\overline{u+v}) = e^{-i\pi E(u,v)}.\overline{\alpha(u)}.\overline{\alpha(v)} = \overline{\alpha(u+v)} ,$$

it is sufficient to define $\alpha(u_i)$'s such that for any element of the basis $\{u_i\}$ we have $\alpha(\overline{u_i}) = \overline{\alpha(u_i)}$.

We can take for the basis $\{u_i\}$ the columns of a matrix Ω of the form defined in theorem (2.3). We then have :

$$\bar{u}_i = u_i \quad \text{if } 1 \leq i \leq n, \qquad \overline{u_{n+j}} = u_j - u_{n+j} \quad \text{if } 1 \leq j \leq \lambda$$

and $\overline{u_{n+k}} = -u_{n+k}$ if $\lambda < k \leq n$.

For $1 \leq i \leq n$, we only need $\alpha(u_i) \in \mathbb{R}$;
for $\lambda < k \leq n$, we need $\overline{\alpha(u_{n+k})} = \alpha(-u_{n+k}) = \alpha(u_{n+k})^{-1}$. In other words, it is sufficient to take $|\alpha(u_{n+k})| = 1$;
finally for $1 \leq j \leq \lambda$, we need :

$$\overline{\alpha(u_{n+j})} = \alpha(u_j - u_{n+j}) = e^{-i\pi E(u_j, u_{n+j})} . \alpha(u_j) . \alpha(u_{n+j})^{-1}$$

or

$$\alpha(u_{n+j}) . \overline{\alpha(u_{n+j})} = e^{-i\pi E(u_j, u_{n+j})} . \alpha(u_j) .$$

For this it is sufficient to take :

$$|\alpha(u_{n+j})| = 1 \text{ and } \alpha(u_j) = e^{i\pi E(u_j, u_{n+j})} \quad (\in \mathbb{R}).$$

It is clear that we can always choose the $\alpha(u_j)$'s in such a manner and this proves the theorem.

(3.6) We will say that (X,E,S) with X, E and S as above (E verifying (3.3)) is a real polarized abelian variety.

Let $D = \begin{pmatrix} d_1 & & 0 \\ & \ddots & \\ 0 & & d_n \end{pmatrix}$ where the d_i's are positive integers such that $d_i | d_{i+1}$. We will say that a polarization is of type D if in a suitable basis (said to be simplectic for E) its matrix is of the form $\begin{pmatrix} 0 & D \\ -D & 0 \end{pmatrix}$.
We will say that E is a principal polarization if $D = I_n$.

To a couple (X,E) formed of an abelian variety and a polarization corresponds a normalized period matrix of the forme (D,Z) with Z a symmetric matrix with imaginary part, ImZ, positive definite. In this case, we have $X = \mathbb{C}^n/[(D,Z)]$, the polarization E is the bilinear form with matrix $\begin{pmatrix} 0 & D \\ -D & 0 \end{pmatrix}$ in the basis of the lattice [(D,Z)] given by the columns of (D,Z) (or equivalently H -defined by (3.2)- has matrix $(\mathrm{Im}Z)^{-1}$ in the canonical basis of $\mathbb{C}^n$). Conversely, if D and Z are matrices of the above type, $X = \mathbb{C}^n/[(D,Z)]$ has a natural structure of polarized abelian variety of type D.

Let $N \in GL_{2n}$ be a symplectic matrix for D (that is, a matrix such that ${}^tN\begin{pmatrix} 0 & D \\ -D & 0 \end{pmatrix}N = \begin{pmatrix} 0 & D \\ -D & 0 \end{pmatrix}$). Write $N = \begin{pmatrix} a & b \\ c & d \end{pmatrix}$ with a, b, c and d in $M_n(\mathbb{Z})$.

Let $\mathcal{H}_n$ be the set of complex symmetric matrices with positive definite imaginary part. N operates on $\mathcal{H}_n$ via :

(3.7) $$N(Z) = (aZ + bD)(cZ + dD)^{-1}D .$$

(3.8) _If_ Z _and_ Z′ _are in_ $\mathcal{H}_n$, _then the polarized abelian varieties defined by_ (D,Z) _and_ (D,Z′) _are isomorphic over_ $\mathbb{C}$ _if and only if there exists_ $N \in GL_{2n}(\mathbb{Z})$, _symplectic for_ D, _such tht_ $N(Z) = Z'$. _The corresponding isomorphism from_ $X' = \mathbb{C}^n/[(D,Z')] \longrightarrow X = \mathbb{C}^n/[(D,Z)]$ _is induced by the linear isomorphism from_ $\mathbb{C}^n$ _to_ $\mathbb{C}^n$ _defined by_ $z \mapsto {}^t(cZ+d)z$ (_recall that_ Z _and_ $(aZ+bD)(cZ+dD)^{-1}D$ _are symmetric_).

4. Real moduli for real principally polarized abelian varieties

Our aim is to build a classifying space for real polarized abelian surfaces. This we will do in §5. But for most of the preliminary results it turns out that if the polarization is principal, then it does note cost any more to work with abelian varieties in arbitrary dimension. In the case of non principal polarization the situation is more complicated and properly belongs to the theory of abelian varieties, so we will only make a few remarks on this case.

It may be useful here to say a few words on what we understand under real moduli. Our interest is in the construction of a space classifying real isomorphy classes of real varieties of a given type. From this point of view, the "classical" approach consisting in looking at the real part of the complex moduli space is note satisfactory. We explain this. The first reason is that a point in the real part of the complex moduli space may not correspond to any real variety. We give an example due to Shimura [Sh_1]. Let a, b and c be irrational numbers with $c > 0$, and consider the matrix :

$$(I_2, Z) = \begin{pmatrix} 1 & 0 & a+ic & b \\ 0 & 1 & b & -a+ic \end{pmatrix}.$$

Let X be the principally polarized abelian surface defined by this period matrix. Let

$$N = \begin{pmatrix} 0 & 1 & & 0 \\ -1 & 0 & & \\ & & 0 & -1 \\ & 0 & 1 & 0 \end{pmatrix},$$

then $N(Z) = \bar{Z}$ (complex conjugate). This means that X is isomorphic to X^σ, its conjugate variety, and the isomorphism is an isomorphism of principally polarized varieties. On the other hand, the choice of a, b and c implies that X is not defined over $\mathbb{R}$ (see (4.1) and (4.2) below).

The second reason is that a real point in the complex moduli space may represent distinct non real isomorphic varieties. For an example, one can consider the elliptic curves E_1 and E_2 defined by $y^2 = x^3+x$ and $y^2 = x^3-x$, respectively. $E_1(\mathbb{R})$ has one component while $E_2(\mathbb{R})$ has two, but they are isomorphic over $\mathbb{C}$. This also provides an example for surfaces, just consider the products $E_i \times E_j$ in order to have an example of a complex surface with at least 3 distinct real models.

There are other reasons why this "classical" approach is not satisfactory, but the two we have given already show that even if we are only interested in classifying real varieties, another approach is necessary.

(4.1) <u>Theorem</u> : <u>Let</u> (X,E) <u>be a principally polarized abelian variety. There exists a real structure</u> S <u>on</u> X <u>such that</u> E <u>is</u> S-<u>real if and only if</u> (X,E) <u>admits a period matrix of the following form</u> :

$$(I_n, (1/2)M + iT),$$

<u>where</u> T <u>is a positive definite real matrix and</u> M <u>is of one of the following forms</u> :

(i) $$M = \begin{pmatrix} I_\lambda & 0 \\ 0 & 0 \end{pmatrix}$$

<u>or</u>

(ii) $$M = \begin{pmatrix} 0 & & 1 & & 0 \\ & & \cdot^{\cdot^{\cdot}} & & \\ 1 & & & & 0 \\ 0 & & & & 0 \end{pmatrix}.$$

<u>Proof</u> : To see that the condition is sufficient, let (I_n,Z) be a period matrix of the above form. The lattice $[(I_n,Z)]$ is then globally invariant under complex conjugation and complex conjugation induces a real structure S on $X = \mathbb{C}^n/[(I,Z)]$. The matrix for the action of S on the lattice, in terms of the basis given by the columns of (I,Z), is $\begin{pmatrix} I & M \\ 0 & -I \end{pmatrix}$. An easy computation shows that, since M is symmetric, the alternating 2-form canonically associated to Z is S-real.

To prove that the condition is necessary, let S be a real structure on X, such that E is S-real. Let $(a_i)_i$ be an S-semi real basis of $H_1(X(\mathbb{C},\mathbb{Z}))$ (see (2.1)) and recall that by definition we have for the n first elements of such a basis $S_*(a_i) = a_i$. The condition (3.3) on E implies that for $1 \leq i,j \leq n$ we have $E(a_i,a_j) = 0$. As a consequence, the matrix of E in this basis is of the form $\begin{pmatrix} 0 & -{}^tA \\ A & B \end{pmatrix}$. Since the polarization is principal $\det A = \pm 1$ and A is invertible. A base change by a matrix of the form $\begin{pmatrix} I_n & 0 \\ 0 & A^{-1} \end{pmatrix}$ transforms the matrix of E into $\begin{pmatrix} 0 & I \\ -I & B' \end{pmatrix}$, where ${}^tB' = -B'$. By the condition on B′, it is easy to find C such that $B' = C - {}^tC$. A base change by a matrix of the form $\begin{pmatrix} I & C \\ 0 & I \end{pmatrix}$ changes the matrix of E into its canonical form $\begin{pmatrix} 0 & -I \\ I & 0 \end{pmatrix}$. Both base changes leave the first n elements fixed. The matrix of S with respect to the first basis was of the form $\begin{pmatrix} I & \Lambda \\ 0 & -I \end{pmatrix}$. In the new basis it is $\begin{pmatrix} I & N \\ 0 & -I \end{pmatrix}$, N with integer coefficients. This implies that $(I,1/2N+iT)$

is a period matrix for (X,E) relative to the new basis. Since this basis is symplectic for E, this implies by the Riemann conditions that N and T are symmetric and that T is positive definite. Base changes given by matrices of the form $\begin{pmatrix} {}^ta^{-1} & b \\ 0 & a \end{pmatrix}$ with ${}^tab$ symmetric are symplectic for E and transform N into ${}^taNa + 2\,{}^tab$. This means that we can reduce N modulo 2, and then, applying Albert's [Al] classification of bilinear forms mod. 2, we can reduce N to one of the forms indicated in the theorem.

We will from now on assume that all the real structures considered on abelian varieties satisfy : $X(\mathbb{R}) \neq \emptyset$. Under this hypothesis, we will say that a matrix $\Omega = (I_n, (1/2)M+iT)$, M as in (4.1) and T real positive definite, represents (X,E,S) if Ω represents (X,E) in the usual sens described at the end of §3 and if (X,S) is real equivalent to $(\mathbb{C}^n/[\Omega],\sigma)$, where σ is induced by complex conjugation on $\mathbb{C}^n$. If there is no ambiguity on the type of polarization, we will simply say that $Z = (1/2)M + iT$ represents (X,E,S).

It remains to determine when two such matrices represent the same isomorphy class. We have :

(4.2) <u>Theorem</u> : <u>Two matrices</u> $Z = (1/2)M + iT$ <u>and</u> $Z' = (1/2)M' + iT'$ (M <u>and</u> M' <u>as in</u> (4.1)) <u>represent isomorphic triples</u> (X,E,S) <u>and</u> (X',E',S') <u>if and only if there exists</u> $a \in GL_n(\mathbb{Z})$ <u>such that</u> ${}^taMa \equiv M'$ (mod. 2) <u>and</u> $T' = {}^taTa$.

<u>Proof</u> : (X,E) and (X',E') are isomorphic over $\mathbb{C}$ if and only if there exists $N \in Sp_{2n}(\mathbb{Z}) = \{N \in GL_{2n}(\mathbb{Z}) \ / \ {}^tN\begin{pmatrix} 0 & -I_N \\ I_n & 0 \end{pmatrix}N = \begin{pmatrix} 0 & -I_n \\ I_n & 0 \end{pmatrix}\}$ such that $N(Z) = Z'$, N(Z) defined by (3.7). Write $\varphi_N \ : X' = \mathbb{C}^n/[(I,Z')] \longrightarrow X = \mathbb{C}^n/[(I,Z)]$ for the morphism induced by $\tilde{\varphi}_N \ : z \longmapsto {}^t(cZ+d)z$ (where $N = \begin{pmatrix} a & b \\ c & d \end{pmatrix}$). S and S' being induced by complex conjugation, φ_N is an isomorphism for the real structures if and only if $\tilde{\varphi}_N$ commutes with complex conjugation, that is : $cZ + d \in GL_n(\mathbb{R})$. Since $\mathrm{Im}Z = T \in GL_n(\mathbb{R})$ is invertible, this is only

possible if $c = 0$.

On the other hand $\begin{pmatrix} a & b \\ 0 & d \end{pmatrix} \in Sp_{2n}(\mathbb{Z})$ if and only if a is invertible, $d = {}^t a^{-1}$ and ${}^t db$ is symmetric. Writing ${}^t a$ in place of a, we can always write $N = \begin{pmatrix} {}^t a & b \\ 0 & a^{-1} \end{pmatrix}$ with $a \in GL_n(\mathbb{Z})$, $b \in M_n(\mathbb{Z})$ and ${}^t a^{-1} b$ symmetric (or equivalently ${}^t a({}^t a^{-1} b) a = ba$ symmetric). We then have $N(Z) = {}^t aZa + ba = Z'$ or ${}^t aTa = T'$ and ${}^t aMa + 2ba = M'$. Since M and M' are symmetric by hypothesis, this proves the theorem.

(4.3) <u>Corollary</u> : <u>If</u> $Z = (1/2)M + iT$ <u>and</u> $Z' = (1/2)M' + iT'$ (M <u>and</u> M' <u>as in</u> (4.1)) <u>correspond to</u> (X',E',S') <u>isomorphic, then</u> rank M = rank M'. <u>If</u> rank M <u>is even then</u> M = M'.

<u>Proof</u> : The first assertion follows immediately from (4.2). The second from the classification of bilinear forms mod. 2 (see for example [Al] theorems 3 and 7). In fact, from this same classification we can say somewhat more. If rank M is odd and if M is of type (i) of (4.1) (respectively type (ii)), then there exists $a \in Gl_n(\mathbb{Z}/2)$ such that $M' \equiv {}^t aMa$ is of type (ii) (respectively of type (i)). Identifying with the corresponding matrix, with 0's and 1's, in $GL_n(\mathbb{Z})$ we can always, if rank M is odd, replace Z by ${}^t aZa = \begin{pmatrix} {}^t a & 0 \\ 0 & a^{-1} \end{pmatrix}(Z)$ and hence assume that M is of type (i).

<u>If</u> rank M <u>is odd we will always make this convention</u>.

(4.4) <u>Definition</u> : <u>we will say that a principally polarized real abelian variety is of type</u> (λ,α) <u>if it has a period matrix of the form</u> : $(I_n, (1/2)M+iT)$ <u>with</u> M <u>as in</u> (4.1) <u>and</u> rank $M = \lambda$ <u>and</u> M <u>of type</u> (α) (α = i <u>or</u> ii). <u>If</u> λ <u>is odd we will always take</u> α = i. <u>If</u> $\lambda = 0$, <u>we will say that the type is</u> (0,0).

With this notation we have :

(4.5) <u>Corollary</u> : <u>The type</u> (λ,α) <u>of a real principally polarized abelian variety</u> (X,E,S) <u>is an invariant of the isomorphy class of</u>

(X,E,S).

In fact, λ has an easy interpretation. If (X,E,S) is of type (λ,α), then the number of connected components of $X(\mathbb{R})$ is $2^{n-\lambda}$ (see (1.9)).

Let $\mathcal{A}^n_{(\lambda,\alpha)}$ be the set of isomorphy classes of real principally polarized abelian varieties of dimension n and type (λ,α).

Let H_n be the set of real symmetric positive definite matrices of rank n.

Let Γ_M (or $\Gamma_{(\lambda,\alpha)}$ if M is of type (λ,α)) be the subgroup of $GL_n(\mathbb{Z})$ of matrices such that ${}^t aMa \equiv M \pmod{2}$. We have :

(4.6) <u>Theorem</u> :

$$\mathcal{A}^n_{(\lambda,\alpha)} = H_n/\Gamma_{(\lambda,\alpha)} ,$$

<u>where</u> $\Gamma_{(\lambda,\alpha)}$ <u>operates via</u> $T \longmapsto {}^t aTa$.

<u>Proof</u> : If λ is even, the theorem follows from (4.2) and (4.3). If λ is odd, the theorem follows from (4.2) the remark made on the odd case and the subsequent convention on α.

The groups $\Gamma_{(\lambda,\alpha)}$ are easy to compute. We have :

$$\Gamma_{(\lambda,i)} = \left\{ \begin{pmatrix} a & b \\ * & * \end{pmatrix} \in GL_n(\mathbb{Z}) \ / \ {}^t aa \equiv I_\lambda \pmod{2} \text{ and } b \equiv 0 \pmod{2} \right\}$$

and, if we write J_λ for the rank λ matrix $\begin{pmatrix} 0 & & 1 \\ & \cdot^{\cdot^{\cdot}} & \\ 1 & & 0 \end{pmatrix}$:

$$\Gamma_{(\lambda,ii)} = \left\{ \begin{pmatrix} a & b \\ * & * \end{pmatrix} \in GL_n(\mathbb{Z}) \ / \ {}^t aJ_\lambda a \equiv J_\lambda \pmod{2} \text{ and } b \equiv 0 \pmod{2} \right\}$$

Note that these groups are of finite index $\leqslant 2^n$ in $GL_n(\mathbb{Z})$. This implies that the $\mathcal{A}^n_{(\lambda,\alpha)}$ are coverings of finite degree of $H_n/GL_n(\mathbb{Z})$.

Using Minkowski's reduction of positive quadratic forms (see for example Weil [We$_2$]), we can describe $H_n/GL_n(\mathbb{Z})$ in the following way :

Let $\mathfrak{R}_n$ be the set of Minkowski reduced matrices, that is the set of real positive definite matrices $A = (a_{ij})$ such that :
(i) $A(x_1,\ldots,x_n) \geqslant a_{ii}$ for every set of integers $(x_1,\ldots,x_n)$ such that $(x_i,\ldots,x_n) = 1$ (relatively prime) ;
(ii) $a_{i,i+1} \geqslant 0$ for $1 \leqslant i \leqslant n-1$.

It is known that $\mathfrak{R}_n$ can be defined by a finite set of the above inequalities and that it is a convex pyramid in $\mathbb{R}^{n(n+1)/2}$. It is also known that there is only a finite number of a's in $GL_n(\mathbb{Z})$ such that ${}^t a\mathfrak{R}_n a \cap \mathfrak{R}_n \neq \emptyset$ and that moreover if this intersection is non empty, then it is again a convex pyramid of smaller dimension contained in the frontier of $\mathfrak{R}_n$. We can thus describe $H_n/GL_n(\mathbb{Z})$ as obtained by identifying a finite number of semi-algebraic sub-sets contained in the faces of the convex pyramid $\mathfrak{R}_n$.

5. Moduli spaces for real polarized abelian surfaces.

For surface the above construction leads to a particularly simple description of the corresponding Moduli spaces. In this case $\mathfrak{R}_2$ is defined by the inequalities $0 < a_{11} \leqslant a_{22}$ and $0 \leqslant 2a_{12} \leqslant a_{11}$ (see for example [We$_2$]). We can also explicitly compute the a's for which ${}^t a\mathfrak{R}_2 a \cap \mathfrak{R}_2 \neq \emptyset$. We represent them symbolically by the following diagram :

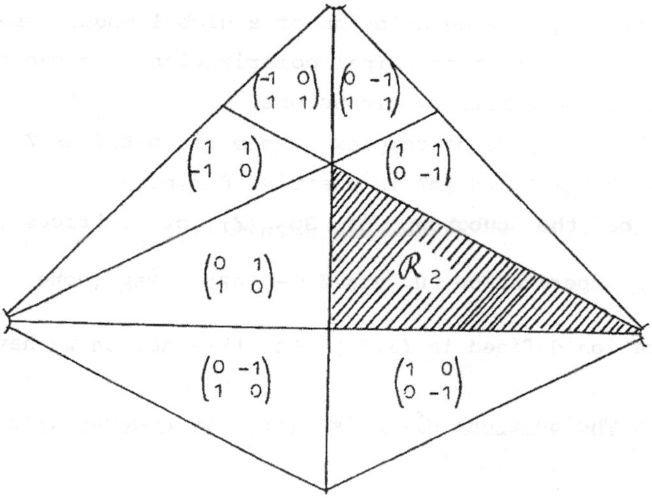

where the horizontal line represents matrices with $y_{12} = 0$; the vertical line matrices with $y_{11} = y_{22}$; the others, respectively, matrices for which $y_{11} = 2y_{12}$, $y_{11} = y_{12}$, $y_{22} = 2y_{12}$, $y_{11} = -2y_{12}$ and $y_{22} = -2y_{12}$.

We note that for each of the a's we have : if x and ${}^t axa$ are both in $\mathfrak{R}_2$, then $x = {}^t axa$. In other words, we have in fact $H_n/GL_n(\mathbb{Z}) = \mathfrak{R}_2$.

For the groups $\Gamma_{(\lambda,\alpha)}$ we have :

$$\Gamma_{(0,0)} = \Gamma_{(2,ii)} = GL_2(\mathbb{Z}) \quad ;$$

$$\Gamma_{(1,i)} = \left\{ N \in GL_2(\mathbb{Z}) \ / \ N \equiv \begin{pmatrix} 1 & 0 \\ * & 1 \end{pmatrix} (\mathrm{mod}.2) \right\} \quad ;$$

$$\Gamma_{(2,i)} = \left\{ N \in GL_2(\mathbb{Z}) \ / \ N = \begin{pmatrix} 1 & 0 \\ 0 & 1 \end{pmatrix} \text{ or } \begin{pmatrix} 0 & 1 \\ 1 & 0 \end{pmatrix} (\mathrm{mod}.2) \right\} .$$

This enables us to write :

$$\mathcal{A}^2_{(0,0)} = \mathcal{A}^2_{(2,ii)} = \mathfrak{R}_2 \quad ;$$

$$\mathcal{A}^2_{(1,i)} = \{ (y_{ij}) \in H_2 \ / \ 0 \leqslant y_{12} \leqslant y_{11} \text{ and } 0 \leqslant 2y_{12} \leqslant y_{22} \} \ ;$$

$$\mathcal{A}^2_{(2,i)} = \{ (y_{ij}) \in H_2 \ / \ 0 < y_{11} \leqslant y_{22} \text{ and } 0 \leqslant y_{12} \leqslant y_{11} \} \ .$$

We can also give a description of a global moduli space. Again, since we are working with principal polarizations, we can do this for abelian varieties of arbitrary dimensions.

Let H_n' be the space of complex symmemtric matrices Z of degree n such that $2\mathrm{Re}Z \in M_n(\mathbb{Z})$ and ImZ is positive definite.

Let Γ_∞ be the subgroup of $Sp_{2n}(\mathbb{Z})$ of matrices of the form $\begin{pmatrix} {}^t a & b \\ 0 & a^{-1} \end{pmatrix}$. Γ_∞ operates on H_n' via $Z \longmapsto {}^t aZa + {}^t ab$ (note that this is just the operation defined in (3.7)). For this action we have :

(5.1) <u>Lemma : The quotient H_n'/Γ_∞ is naturally endowed with the struc-</u>

ture of a real analytic space.

Proof : First recall that $Sp_{2n}(\mathbb{Z})$ operates properly discontinuously on the Siegel upper half space $\mathcal{H}_n$ and hence so does $\Gamma_\infty \subset Sp_{2n}(\mathbb{Z})$. This implies that Γ_∞ operates properly discontinuously on H_n', and the assertion follows from a well known theorem of Cartan [Ca].

Let $\mathcal{A}_{\mathbb{R}}^n$ be the set of isomorphism classes of triples (X,E,S) where X is an abelian variety of dimension n, S a real structure on X and E an S-real principal polarization on X.

We have :

(5.2) *Proposition* :

$$\mathcal{A}_{\mathbb{R}}^n = H_n' / \Gamma_\infty$$

and $\mathcal{A}_{\mathbb{R}}^n$ *has* n+1 + (n-1)/2 *if* n *is odd or* n+1 + n/2 *if* n *is even, connected components, which can be identified with the spaces* $\mathcal{A}_{(\lambda,\alpha)}^n$.

Proof : By the same argument as in the proof of (4.1), each Z in H_n' corresponds to a real principally polarized abelian variety (X,E,S) of dimension n and conversely.

Z and Z' correspond to $\mathbb{C}$-isomorphic (X,E) and (X',E') if and only if there exists $N \in SP_{2n}(\mathbb{Z})$ such that N(Z) = Z'. As we have noted in the proof of (4.2), such an isomorphism is real (that is compatible with S and S') if and only if N is of the form $\begin{pmatrix} {}^t a & b \\ 0 & a^{-1} \end{pmatrix}$. In other words, if and only if $N \in \Gamma_\infty$.

Let $Z \in H_n'$ and ReZ = ½M, M of type (λ,α) (see (4.4)). Call $\Gamma_\infty(Z)$ the orbit of Z with respect to the action of Γ_∞ and $\Gamma_{(\lambda,\alpha)}(Z)$ the orbit of Im(Z) with respect to the action of $\Gamma_{(\lambda,\alpha)}$. Then it is easy to see that $\Gamma_\infty(Z) \cap (½M + iH_n) = ½M + i\Gamma_{(\lambda,\alpha)}(Z)$. To end the proof, we only have to note that by (4.3) and the comment that follows the proof of (4.3), each connected component of $\mathcal{A}_{\mathbb{R}}^n$ corresponds to exactly one of the types (λ,α) (recall that by construction, H_n being connected, each of the $\mathcal{A}^{\eta}_{(\lambda,\alpha)}$ are connected).

(5.3) Remark : The case of an arbitrary polarization, even for surfaces, is not at all as simple as the preceding one. We can describe roughly the situation as follows :

If the polarization is of type $D = \begin{pmatrix} d_1 & 0 \\ 0 & d_2 \end{pmatrix}$ with d_2/d_1 odd, then things are not very different from what happens in the principle case. To obtain a theorem like (4.1), one must consider the matrices : $M = 0$, $\begin{pmatrix} 1 & 0 \\ 0 & 0 \end{pmatrix}$, I_2 and $\begin{pmatrix} 0 & d_2/d_1 \\ 1 & 0 \end{pmatrix}$. If d_2/d_1 is divisible by 2 but not by 4, one still has a satisfying result. One has only to consider an additional case, the case $M = \begin{pmatrix} 0 & 0 \\ 0 & 1 \end{pmatrix}$.

If d_2/d_1 is divisible by 4, then the results are only easy to describe if the invariant λ corresponding to the real structure is ≤ 1.

In all the above cases, the groups Γ_M are easy to compute and we obtain Moduli spaces of the form : H_n/Γ_M.

Bibliographical Notes :

The results of §1 and §2 are due to Comessatti [Co_2]. The presentation given here is taken from [Si_1].

(3.3) and (4.1) are inspired by Shimura [Sh_2] and the example at the biginning of §4 comes from Shimura [Sh_1].

V. REAL RULED SURFACES

1. General properties of real fibred surfaces

Let X_o and B_o be respectively a surface and a curve over $\mathbb{R}$.

(1.1) We will say that X_o is fibred over B_o if there exists a projective morphism $p : X_o \longrightarrow B_o$ such that if η is a generic point of B_o then $p^{-1}(\eta)$ is a smooth curve over the field $\mathbb{R}(\eta)$. <u>We will say that</u> p <u>is a real fibration</u>.

We can reformulate this in terms of definition I.(1.15) :

(1.2) Let (X,S) be a real surface and let (B,S') be a real curve. Let $p : X \longrightarrow B$ be a fibration of X over B, as a complex surface. Then p comes from a real fibration for the corresponding real models of (X,S) and (B,S') if and only if $p \circ S = S' \circ p$ (the "if and only if" follows immediately from I.(1.4)).

(1.3) Since the morphism p of (1.1) is projective, we may, performing a Stein factorization if necessary (that is replacing B by a finite cover B'), assume that the fibres of p are connected. <u>We will always make this assumption</u>.

We will say that the surface X_o in (1.1) (or the surface X in (1.2)) is <u>ruled</u> if $p^{-1}(\eta)$ is of genus 0 and that it is <u>elliptic</u> if $p^{-1}(\eta)$ is of genus 1.

(1.4) Let $p : X \longrightarrow B$ and $p' : X' \longrightarrow B$ be two fibred surfaces over the same curve B. We will say that a morphism $f : X \longrightarrow X'$ is a <u>B-morphism</u> if it commutes with the fibration.

(1.5) Definition : Let X_0 be a smooth irreducible surface over $\mathbb{R}$ and let $p : X_0 \longrightarrow B_o$ be a real fibration. We will say that X_0 is B_o-minimal if no fibre contains an exceptional curve of the first kind in the sense of definition II.(6.3).

Of course we have the corresponding definition for complex fibrations. $p : X \longrightarrow B$ is B-minimal if no fibre of p contains an exceptional curve of the first kind in the ordinary sense.

Relating the two we have :

(1.6) Theorem (Manin) : Let X_0 be a smooth projective surface over $\mathbb{R}$ and let $p : X_0 \longrightarrow B_o$ be a real fibration (we assume -see (1.3)- that the fibres are connected). Let X_η be a generic fibre.

(i) If the genus of X_η over $\mathbb{R}(\eta)$ is ≥ 1, then X_0 is B_o-minimal if and only if $X = X_0 \times_{\mathbb{R}} \mathbb{C}$ is B-minimal (where $B = B_o \times_{\mathbb{R}} \mathbb{C}$).

(ii) If the genus of X_η is zero and if X_0 is B_o-minimal, then the fibres of $p_{\mathbb{C}} : X_0 \times_{\mathbb{R}} \mathbb{C} \longrightarrow B = B_o \times_{\mathbb{R}} \mathbb{C}$ which contain an exceptional curve of the first kind are of the form $L + L^S$, where L is exceptional of the first kind and $L.L^S = 1$.

Proof : Assume that there is an exceptional curve of the first kind L in a fibre $F = p_{\mathbb{C}}^{-1}(b)$. Since the fibration is real, L^S must also be in a fibre . If X_0 is B_o-minimal, then $L^S.\ L > 0$ and $L^S \neq 1$. In particular, L^S and L are in the same fibre.

This implies $F^S = F$ (in other words the fibre is over a real point of B) and we can write :

$$F = n(L+L^S) + \Sigma\, n_j C_j$$

with $n > 0$, $S(\Sigma\, n_j C_j) = \Sigma\, n_j C_j$ and for all j, $C_j \neq L$ and L^S.

Since F is a fibre, $(F.L) = 0$. But $(F.L) = n(L+L^S).L + \Sigma n_j C_j.L$ and since $L^2 = -1$ and $L.L^S > 0$, this implies that $C_j.L = 0$ for all j.

The same argument with L^S in place of L implies that $C_j.L^S = 0$.

Since the fibre is connected, this shows that for all j, $n_j = 0$.

Hence $F = n(L+L^S)$.

If X is a ruled surface (genus of X_η is 0) then it is well known (see for example [Gr&Ha] p. 565) that X cannot have multiple fibres. Hence $n = 1$. We have noted that $L.L^S > 0$. If $L.L^S > 1$ then the B-minimal model of X over $\mathbb{C}$ would have a singular fibre which is impossible. This proves (ii).

If the genus of the generic fibre is ≥ 1, then the semi-continuity theorem implies that we must have $p_a(F) \geq 1$.

Now since $p_a(L) = 0$ and $L^2 = -1$, we must have by the adjunction formula $K.L = -1$ (where K is a canonical divisor) and also for the same reasons $K.L^S = -1$. Noting that $(L+L^S)^2 = 0$, the adjunction formula implies that $p_a(n(L+L^S)) \leq 0$. In other words, we cannot have a fibre of this type and (i) is proved.

We now turn to the problem of existence of fibrations of genus 0 or 1. Our first result in this direction is due to Enriques (cf. Manin [Ma_2] theorem 1.2 p. 141) :

(1.7) <u>Theorem</u> (Enriques) : <u>Let</u> X_o <u>be a rational surface over</u> $\mathbb{R}$. <u>Then there exists a smooth projective surface</u> X'_o <u>over</u> $\mathbb{R}$, $\mathbb{R}$-<u>birationally equivalent to</u> X_o, <u>a smooth curve of genus</u> 0 <u>over</u> $\mathbb{R}$, B_o <u>and a real fibration</u> $p : X'_o \longrightarrow B_o$ <u>of genus</u> 0 <u>or</u> 1 (<u>that is the generic fibre</u> X'_η <u>is of genus</u> 0 <u>or</u> 1 <u>over</u> $\mathbb{R}(\eta)$).

<u>Sketch of proof</u> (we will sharpen this result in VI.§2 and §4) : Let K be a canonical divisor on X. By a classical result (see [B&P&V] p. 190), we have for any divisor D and for m large enough :

$$h^0(mK + D) = \dim H^0(X,\mathcal{O}(mK + D)) = 0.$$

Now let D be a very ample divisor, hence in particular such that $h^0(D) \geq 3$. By the above we can find an n such that :

$$h^0(nK + D) \geq 2 ,$$

$$h^0((n+1)K + D) \leq 1 .$$

Let f_0 and f_1 be linearly independant elements of $H^0(X,\mathcal{O}(nK + D))$ (by I.(4.3) we can always choose these to be real) and consider the rational map : $X \longrightarrow \mathbb{P}^1$ defined by f_0 and f_1. Blowing up points of indeterminacy of this map if any (which are real points or pairs of complex conjugate points) and performing a Stein factorisation if necessary we obtain a morphism

$$p : X' \longrightarrow B,$$

where X' is a real surface, birationally equivalent to X by construction, B a real curve which is smooth because it maps onto $\mathbb{P}^1$ and of genus 0 because X and X' are rational.

p is also real and the genus of the fibration is $\leq p_a(nK + D)$. We compute this.

Since $h^0(nK + D) \geq 2$, there exists an effective divisor C in the linear system $|nK + D|$. We apply Riemann-Roch to -C :

$$h^0(-C) + h^0(K + C) \geq (1/2)(C^2 + K.C) + 1.$$

But $h^0(-C) = 0$ because C is effective and $h^0(K + C) = h^0((n+1)K + D) \leq 1$. Hence :

$$1 \geq (1/2)(C^2 + K.C) + 1 = p_a(C)$$

by the adjunction formula. Hence the claim.

If X is not rational and ruled we have :

(1.8) <u>Proposition</u> : <u>If</u> X_0 <u>is a non rational surface over</u> $\mathbb{R}$, <u>such that</u> $X_0(\mathbb{R}) \neq \emptyset$ <u>and such that</u> $X = X_0 \times_{\mathbb{R}} \mathbb{C}$ <u>is ruled</u>, <u>then</u> X_0 <u>is ruled over an</u> $\mathbb{R}$-<u>curve</u> B_o.

<u>Proof</u> : If X is non rational and ruled, then the irregularity q of X is > 0, and the albanese map $a_X : X \longrightarrow Alb(X)$ defines a ruling on X (cf. for example [Sha] Chap. IV or [Be] p. 85-86). But by a well known and easy result (see IV. (2.4)) the condition $X(\mathbb{R}) \neq \emptyset$ implies that the albanese map is real. This proves (1.8).

For elliptic surfaces we do not have a result of this type. To find a counter example, we consider a surface of the form $F \times F^\sigma$, where F is an elliptic curve and F^σ is the conjugate curve (see I.(1.1)). This is obviously an elliptic surface and it has a real structure defined by the involution $(x,y) \longmapsto (\sigma y, \sigma x)$. We are going to show that we can choose F in such a way that there is no real fibration of genus 1. For this, we recall that an elliptic curve is defined over $\mathbb{R}$ if an only if it is isogenous to a curve defined by a period matrix of the form [1,ai], $a \in \mathbb{R}$ and $i^2 = -1$ (cf. IV.(2.3)). This implies that if F is defined by a period matrix of the form [1,b+i], with b irrational, then F is not isogenous to a curve defined over $\mathbb{R}$. Choose F in this way and assume that there exists a real fibration $F \times F^\sigma \longrightarrow E$. $F \times F^\sigma$ must be isogenous to $E \times E'$, where E' is another elliptic curve. From the unicity, up to isogeny, of the decomposition of an abelian variety into a product, this implies that E (resp. E') must be isogenous to F or F^σ. By the choice of F, this implies that E cannot be real, contradicting the assumption on the existence of a real fibration.

In spite of this negative result there are some cases where an elliptic fibration does go down to one over $\mathbb{R}$. For example :

(1.9) (i) If the real surface X is such that the Kodaira dimension, Kod(X) = 1 (that is if X is a so-called properly elliptic surface). The reason for this comes from the fact that the fibration is defined by the map associated to a multiple of the canonical divisor and we can apply I.(4.3).

(ii) If the surface is hyperelliptic then the fibration can be defined by the Albanese map (see [Sha] Chap. IV. §2 or [Be] p. 86) and we can apply the same argument as in the proof of (1.8).
(iii) If X is an Enriques surface and $\chi X(\mathbb{R}) \leq 8$ (see [Si_3], p. 478).

2. Birational classification of real ruled surfaces

The results and methods of this part are generalizations of the ones of Iskovskih [Is_1] and Comessatti [Co_1].

Let X_0 be a smooth irreducible surface over $\mathbb{R}$, with a real ruling $\pi : X_0 \longrightarrow B_o$ over a smooth curve B_o. We assume that X_0 is B_o-minimal and that $X_0(\mathbb{R}) \neq \emptyset$ (which implies $B_o(\mathbb{R}) \neq \emptyset$).

Extending scalars to $\mathbb{C}$, we get a ruling $\pi : X \longrightarrow B$. By Tsen's theorem (see for example [Sha] p. 24), this ruling has a section. The generic fibre $\pi^{-1}(\eta)$ is thus isomorphic to $\mathbb{P}^1$ over $\mathbb{C}(\eta) = \mathbb{C}(B)$ the function field of B.

From this it follows that birationally classifying real surfaces over B is equivalent to classifying Brauer-Severi curves over the field $\mathbb{R}(B)$ (that is, curves over $\mathbb{R}(B)$ isomorphic to $\mathbb{P}^1_{\bar{K}}$ over $\bar{K}$, the algebraic closure of $\mathbb{R}(B)$). In particular, if the fibration $\pi : X_0 \longrightarrow B_o$ has a section, then X is birationally equivalent to $\mathbb{P}^1_{\mathbb{R}} \times B$ over $\mathbb{R}$.

As is well known, Brauer-Severi varieties over a field K are classified, up to K-birational isomorphism, by $H^1(G(\bar{K}|K), PGL_n(\bar{K}))$ where $\bar{K}$ is the algebraic closure of K, n is the dimension of X and $G(\bar{K}|K)$ is the Galois group (see for example Serre [Se_2] p. 166).

In our case, $K' = \mathbb{C}(\eta)$ being a c_1-field, we have in fact :

$$H^1(G(\bar{K}|K), PGL_n(\bar{K})) = H^1(G(K'|K), PGL_n(K')) .$$

For commodity we will denote $H^1(PGL_n(K'))$ this group. For the computation we first note that, since $B = B_o \times_{\mathbb{R}} \mathbb{C}$, we have $K' = \mathbb{C}(B) = \mathbb{R}(B_o) \otimes_{\mathbb{R}} \mathbb{C} = K[\sqrt{-1}]$ and hence $G(K'|K) = G(\mathbb{C}|\mathbb{R}) = G$.

Next we consider the exact sequence :

$$0 \longrightarrow K'^* \longrightarrow GL_2(K') \longrightarrow PGL_1(K') \longrightarrow 0 .$$

By Hilbert's theorem 90, we have $H^1(G, GL_2(K')) = 0$. Also because $2 = [K':K]$, the map $H^1(PGL_1(K')) \longrightarrow H^2(G,K'^*)$ is surjective (see Serre [Se$_2$] Chap. X proposition 8 and lemme 1 p. 166). In other words, $H^1(PGL_1(K')) \simeq H^2(G,K'^*)$.

Now by definition $H^2(K'^*) = K^*/N_{K'|K}$ and because $K' = K[\sqrt{-1}]$ we have for the norm subgroup : $N_{K'|K} = \{f = g^2 + h^2 \ / \ g, h \in K\}$. Hence :

$$(2.1) \qquad H^1(PGL_1(K')) = K^*/\{g^2+h^2 \ / \ g \text{ and } h \text{ in } K\} .$$

To refine this computation and to obtain an explicit set of representatives, we are going to use a theorem of Witt (see [Wit] or Knebusch [Kn]).

The first consequence of this theorem is the following : let $f \in K^* = \mathbb{R}(B)^*$ and call f_∞ the set of points in $B(\mathbb{R})$ where f is not defined. Then :

$$(2.2) \qquad N_{K'|K} = \{f \in \mathbb{R}(B) \ / \ f(x) \geq 0 \text{ for all } x \in B(\mathbb{R}) \backslash f_\infty\} .$$

The second is that if $(B_i)_{1 \leq i \leq n}$ is the set of components of $B(\mathbb{R})$, then there exists for each i a function $g_i \in \mathbb{R}(B)^*$ such that :

$$(2.3) \qquad \begin{array}{lll} g_i(x) < 0 & \text{for} & x \in B_i \\ g_i(x) > 0 & \text{for} & x \in B_j,\ j \neq i . \end{array}$$

The third is that if $(x_j)_{j \in J}$ is a set of distinct points of $B(\mathbb{R})$ such that for each i, $(x_j)_J \cap B_i$ has even order, then there exists an f in $\mathbb{R}(B)^*$ such that :

$$(2.4) \qquad (f)_{|B(\mathbb{R})} = D = \sum_{j\in J} x_j \ ,$$

where $(f)_{|B(\mathbb{R})}$ is the restriction of the divisor (f) to $B(\mathbb{R})$ (the divisor of real zeros and poles of f, or here, by the choice of D, the divisor of real zeros of f). We note that since the x_j's are distinct f changes sign at precisely the points x_j's.

Recalling that, if a function h has a zero or a pole of even order at a point $x \in B(\mathbb{R})$, then h does not change sign in a neighbourhood of x in $B(\mathbb{R})$, we conclude from (2.1), (2.2), (2.3) and (2.4) that every element in $H^2(K'^*)$ has a representative of the form :

$$(2.5) \qquad \left(\prod_{i\in P} g_i\right).f \quad ,$$

where P is a subset of $\{1,\ldots,n\}$, the g_i's are as defined in (2.3) and f as in (2.4) for some set of points $\{x_j\}$.

Recalling that $H^2(K'^*) \cong H^1(PGL_1(K'))$ and that this last group classifies Brauer-Severi curves over $K = \mathbb{R}(B)$, we have :

(2.6) <u>Theorem</u> : <u>Every Brauer-Severi curve over</u> $\mathbb{R}(B)$ <u>has up to an</u> $\mathbb{R}(B)$ <u>isomorphism a representation with affine equation of the form</u> :

$$x^2 + y^2 = \left(\prod_{i\in P} g_i\right).f,$$

$\left(\prod g_i\right).f$ <u>as constructed in</u> (2.5).

(2.7) <u>Corollary</u> ; <u>Let</u> X <u>be a smooth irreducible real surface with a real ruling</u> $\pi : X \longrightarrow B$ <u>over a smooth real curve</u> B. <u>Then</u> X <u>is</u> $\mathbb{R}$-<u>birationaly equivalent</u> (<u>and even</u> B-<u>birationally</u>) <u>to a surface defined in some affine open subset of</u> $\mathbb{A}^2 \times B$ <u>by an equation of the form</u> :

$$x^2 + y^2 = \left(\prod_{i \in P} g_i\right).f,$$

$\left(\prod g_i\right).f$ *as in* (2.5).

Proof : This follows immediately from (2.6) and the reduction of the classification problem made at the beginning of this §.

(2.7) does not completely settle the birational classification. To refine this classification, we must distinguish between irrational and rational surfaces. We do the rational case is the next Chapter. For irrational ruled surfaces we have :

(2.8) *Proposition* : *Let* $\pi : X \longrightarrow B$ *and* $\pi' : X' \longrightarrow B'$ *be two real ruled surfaces over smooth irreducible real curves* B *and* B'. *If* genus $B > 0$, *then for every real birational map* $f : X \longrightarrow X'$ *there exists a real isomorphism* $g : B \longrightarrow B'$ *such that the diagram*

$$\begin{array}{ccc} X & \xrightarrow{\ f\ } & X' \\ {\scriptstyle\pi}\big\downarrow & & \big\downarrow{\scriptstyle\pi'} \\ B & \xrightarrow{\ g\ } & B' \end{array}$$

commutes.

Proof : We first recall that the irregularity of a surface is a birational invariant. In the case of a ruled surface, this is just the genus of the base curve. In particular we have genus B = genus $B' > 0$.

By II.(6.4), we can find a smooth real surfaces X'' and finite sequences of monoidal transformations σ_1 and σ_2 factorizing f

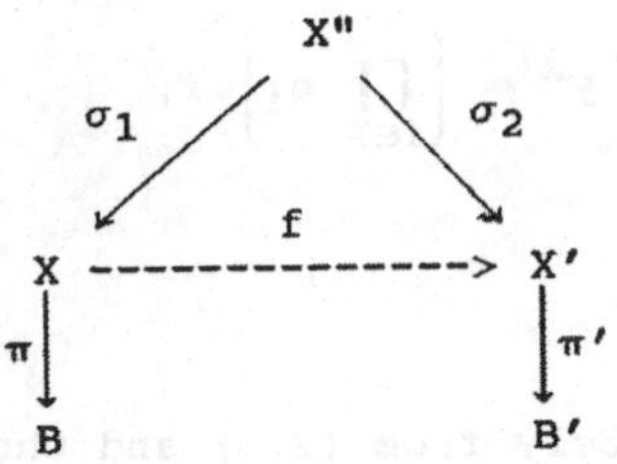

On X" we have two fibrations, respectively on B and B'. Since B and B' are irrational, any rational curve on X" must be contained in a fibre, and this, for both fibrations (if not, we would have a non constant map from a rational curve to B or B'). Now let F be a fibre of $\pi : X \longrightarrow B$. $\sigma_1^{-1}(F)$, being connected and composed of rational curves, must be contained in a unique fibre of $X" \longrightarrow B'$. We obtain in this way a well defined morphism $g : B \longrightarrow B'$. Working with f^{-1} in place of f it is easy to prove that g is an isomorphism.

To formulate the corollary of (2.8) ending the classification of irrational ruled surfaces, we need a few notations.

Let $\{g_i\}$ be functions on B as defined in (2.3) and let P be a subset of $\{1,\dots,n\}$, $n = {}^{\#}B(\mathbb{R})$ the number of connected components of $B(\mathbb{R})$. We call P^+ (resp. P^-) the set of components of $B(\mathbb{R})$ over wich $\prod_{i\in P} g_i > 0$ (resp. $\prod_{i\in P} g_i < 0$). We also recall for the intelligibility of (2.9) that by definition, if f is as in (2.4), then f has only simple zeros and no poles in $B(\mathbb{R})$.

(2.9) Corollary : Let $X \longrightarrow B$ and $X' \longrightarrow B$ be two ruled surfaces over an irrational real curve B. If X (resp. X') is birationally equivalent to the affine surface defined by $x^2+ y^2 = \left(\prod_{i\in P} g_i\right).f$ (resp. $x^2+ y^2 = \left(\prod_{i\in P'} g_i\right).f'$) (notations as in (2.7)), then X is $\mathbb{R}$-birationally isomorphic to X' if and only if there exists a real automorphism φ of B such that :

(i) $\varphi(P^+) = P'^+$ (resp. $\varphi(P^-) = P'^-$) (see above) ;

(ii) $\varphi(\{x_i\}) = \{x_i'\}$, where $\{x_i\}$ (resp. $\{x_i'\}$) is the set of real zeros of f (resp. f').

3. The real part of a ruled surface

We first note the following :

(3.1) Lemma : Assume genus B > 0 and let X be the smooth real relatively minimal (see II.(6.6)) projective completion of the affine surface defined by the equation given in (2.7). Then the fibres over the zeros of f are of the form $L+L^S$ where $L^2 = -1$ and $L.L^S = 1$ and these are the only exceptional curves of the first kind on X.

Proof : That the fibres over the zeros of f are of the above type is obvious. On the other hand, if E is exceptional, then since B is irrational, E must be in a fibre. Also, by the proof of (1.6), since X is real relatively minimal, E must be in a fibre over a real point. But then, by definition, E must be in a fibre over a zero of f.

(3.2) Lemma : If X is an in (3.1), then the real part X(ℝ) of X is topologically the disjoint union of t tori and s spheres with $t \leqslant {}^{\#}B(\mathbb{R})$, the number of connected components of B(ℝ), and s = 1/2(number of real zeros of f).

Proof : Immediate from the fact that $X(\mathbb{R}) \subset \mathbb{A}^2 \times B$ and that the equation is given in (2.7).

(3.3.) Remark : We note that the conclusions of (3.2) still hold even if genus B = 0.

(3.4) Theorem : Let X be a real projective ruled surface over a real curve B. If X is real relatively minimal and if X(ℝ) ≠ ∅, then X(ℝ) is topologically the disjoint union of t tori T_1, k Klein bottles U_1 and s <u>spheres with</u> $t+k \leq {}^{\#}B(\mathbb{R})$ (<u>the number of connected components of</u> $B(\mathbb{R})$) <u>and</u> $2s = \text{rank } NS(X)-2$ (<u>where</u> $NS(X)$ <u>is the Néron-Severi group of</u> X).

To prove (3.4) we will need a definition and a lemma.

Let x be a real point of X not contained in any $L \cap L^S$ for an exceptional curve of the first kind, L. If we blow up x, we obtain a real birational transformation. In the surface X thus obtained, the strict transform of the fibre F_x through x, thas self intersection -1 and of course is real. Hence we can blow down this fibre over $\mathbb{R}$. We obtain in this way a real surface X′ and a real birational transformation $X \longrightarrow X'$ which we note elm_x.

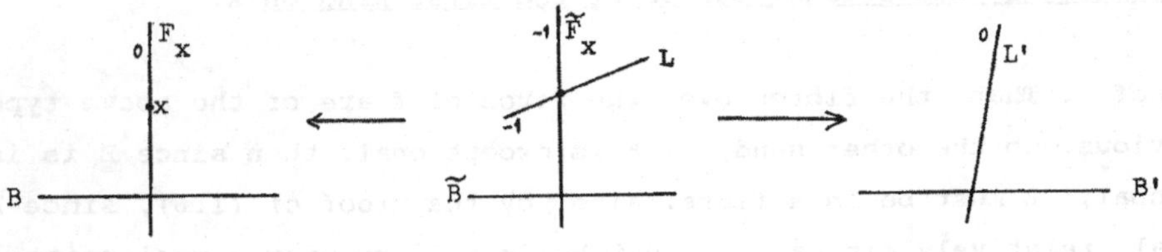

If y is a non real point not contained in an S-invariant fibre, we can blow up, over the reals, simultaneously y and S(y) and then, again over $\mathbb{R}$, blow dow the corresponding fibres. In this way we obtain again a real birational transformation. We will denote $elm_{y,S(y)}$ such a transformation.

(3.5) <u>Definition</u> : <u>Let</u> X <u>and</u> X′ <u>be real ruled surfaces over a real curve</u> B. <u>We will say that a birational transformation</u> $X \longrightarrow X'$ <u>is a real elementary transformation if it is of the form</u> elm_x <u>or</u> $elm_{y,S(y)}$ <u>as defined above</u>.

(3.6) <u>Lemma</u> : <u>Let</u> X <u>and</u> X′ <u>be projective real ruled surfaces over an irrational real curve</u> B. <u>If</u> X <u>and</u> X′ <u>are both real relatively minimal</u>,

<u>then every birational transformation</u> X ⟶ X′ <u>is a product of real elementary transformations</u>.

<u>Proof</u> : By (2.8) we may always assumes that the birational transformation X ⟶ X′ induces the identity on B.

Checking the possibilities for the self-intersection of the blown up fibres, (3.6) is an easy consequence of II.(6.4).

<u>Proof of</u> (3.4) <u>in the case</u> B <u>is irrational</u> (we will prove (3.4) in the case B is rational in the next Chapter) : If B is irrational, everything except the last assertion of (3.4) is an immediate consequence of (3.6) and II.(6.9). To prove the last assertion, let $L_i+S(L_i)$ be the singular fibres. Blowing down the L_i's, we obtain by (3.1) a complex relatively minimal ruled surface X′. It is well known that in this case rank NS(X′) = 2 and is generated by the class of a fibre and the class of a section. Since, by (3.1) and (3.2), the number of spheres in X(ℝ) is half the number of L_i's, the last assertion of (3.4) easily follows from the fact that $NS(X) = NS(X') \oplus_i l_i\mathbb{Z}$, where l_i is the class of L_i (see [Ha] V.3.2).

4. **<u>An example of a non Galois-Maximal real surface</u>**

We end this Chapter by showing that a real ruled surface is in general not Galois-Maximal (see I.(3.11)).

For this let B be a smooth projective real curve of genus ⩾ 1 such that B(ℝ) has strictly more than 1 component. Let $B_1,\dots,B_r$ be the connected components and let $g_1,\dots,g_r$ be corresponding functions as defined in (2.3) (that is, $g_i < 0$ on B_i and $g_i > 0$ on B_j for $j \neq i$).

Now let X be the smooth projective real relatively minimal completion of the surface defined in an affine open set of $\mathbb{A}^2\times B$ by :

$$x^2+ y^2 = - \prod_{i\in P} g_i(b) \qquad b \in B$$

where $P \subsetneq \{1,\ldots,r\}$. Such a surface has $\mathrm{Card}(P) = p < r$ connected components each homeomorphic to a torus T_1. We hence have $\dim H^*(X(\mathbb{R}),\mathbb{Z}/2) = 4p$.

On the other hand, since the surface is ruled, we have

$$H^1(X(\mathbb{C}),\mathbb{Z}) \cong H^1(B(\mathbb{C}),\mathbb{Z}) \ .$$

The ruling being real this isomorphism is compatible with Galois action. By standard computations (see Knebusch [Kn] or [Si_1])

$$\dim H^1(G,H^1(B(\mathbb{C}),\mathbb{Z})) = r-1 \ .$$

Hence by I.(3.9), since $\dim H^1(B(\mathbb{C}),\mathbb{Z}/2) = \mathrm{rank}\, H^1(B(\mathbb{C}),\mathbb{Z}) = 2g$,

$$\dim H^1(G,H^1(B(\mathbb{C}),\mathbb{Z}/2) = 2r - 2 \ .$$

Using Poincaré duality and the fact that $\dim H^1(G,H^0(X)) = \dim H^1(G,H^4(X)) = 1$ we find :

$$\dim H^1(G,H^1(X(\mathbb{C}),\mathbb{Z}/2)) \geqslant 4r - 2 \ .$$

Since $p < r$, we see that the surface X cannot be Galois-Maximal.

<u>Bibliographical Notes</u> :

§1 is inspired by Manin [Ma_1] and [Ma_2].

Iskovskih has proved results similar to the results in §2 in the case of rational ruled surfaces (see [Is_1] and [Is_2]).

The example of §4 is taken form [Si_3].

VI. **RATIONAL REAL SURFACES**

1. Real forms of complex relatively minimal rational surfaces

Recall that a complex relatively minimal rational surface is isomorphic (biregularly) either to $\mathbb{P}^2$ or to a Hirzebruch surface F_n (denoted Σ_n in [B&P&V] and X_n in [Ha]). If n = 0, then X is a quadric in $\mathbb{P}^3$. We will study these in §5. The surface F_n is isomorphic to the projective bundle associated to the rank 2 vector bundle corresponding to $\mathcal{O}_{\mathbb{P}^1} \oplus \mathcal{O}_{\mathbb{P}^1}(n)$. The natural real structure of $\mathbb{P}^1$ induces canonically a real structure on F_n. When we refer, in the sequel, to the surface F_n, we will always assume that it is endowed with this real structure.

We start by the well known :

(1.1) Proposition : If X is a real surface complex isomorphic (biregularly) to $\mathbb{P}^2$, then X is real isomorphic to $\mathbb{P}^2$.

Proof : The Brauer group of $\mathbb{R}$, classifying Brauer-Severi varieties over $\mathbb{R}$ (that is, varieties isomorphic over $\mathbb{C}$ to some $\mathbb{P}^n$), is $\mathbb{Z}/2$ (see [Se_1] p. 170). But the non trivial Brauer-Severi variety is the conic $x^2 + y^2 + z^2 = 0$, which has dimension 1.

(1.2) Proposition (Manin) : If X is a real surface complex isomorphic to a Hirzebruch surface F_n (n > 0) then :

(i) If $X(\mathbb{R}) \neq \emptyset$, then X is real isomorphic to F_n ;

(ii) If $X(\mathbb{R}) = \emptyset$, then n is even ;

(iii) If n is odd, X is real isomorphic to F_n.

Proof : We recall that a surface F_n (n > 0) is characterized by the fact that it is ruled and contains a unique section B of the ruling with $B^2 = -n$. This is the only curve in X with negative self-intersection.

By the unicity of B, we must have $S(B) = B$. If $X(\mathbb{R}) \neq \emptyset$, let x be a real point and F_x the fibre through x. Since x is real we must have $S(F_x) = F_x$. Hence the unique point in $B \cap F_x$ is a real point. This means that B is real-isomorphic to $\mathbb{P}^1$. But in this case the ruling is real and this proves (i).

Since $B^2 = -n$, (ii) is a direct consequence of III.(4.2).

(i) and (ii) prove (iii).

Concerning the real part of F_n we have :

(1.3) <u>Proposition</u> : <u>If</u> X <u>is a real surface isomorphic to</u> F_n $(n > 0)$ <u>we have</u> :

$X(\mathbb{R}) = T_1$ (<u>torus</u>) <u>if</u> n <u>is even</u> ;

$X(\mathbb{R}) = U_1$ (<u>Klein bottle</u>) <u>if</u> n <u>is odd</u>.

<u>Proof</u> : There are several ways to prove this. One way is to note that $\mathrm{Pic}(X) = \mathbb{Z}^2$, generated by the class of the section B and the class of a fibre, both S-invariant.

This means that $H^1(G,\mathrm{Pic}(X)) = 0$ and $H^2(G,\mathrm{Pic}(X)) = (\mathbb{Z}/2)^2$. By III.(3.3) this proves that X is connected and by III.(4.1.1), that $H^1(X(\mathbb{R}),\mathbb{Z}/2) = (\mathbb{Z}/2)^2$. To conclude the proof we only need to use III.(4.1) or III.(4.2).

2. <u>Manin's classification of rational surfaces.</u>

Although the general method we are going to use here is originally due to Enriques and Castelnuovo (the method of adjunction), the precise statements we are going to establish in this part are all due to Manin [Ma_1], hence the title.

We will say that a divisor is <u>real irreducible</u> if either it is real $(D = S(D))$ and irreducible over $\mathbb{C}$ (in which case we will say that it is a real irreducible curve), or if it is of the forme $D' + S(D')$ with D' irreducible over $\mathbb{C}$ and $D' \neq S(D')$.

(2.1) <u>Lemma</u> : <u>If a smooth rational real surface</u> (<u>with</u> $X(\mathbb{R}) \neq \emptyset$) <u>contains a real irreducible effective divisor</u> C <u>such that</u> $p_a(C) \leq 0$ <u>and</u> $C^2 \geq 0$, <u>then</u> X <u>is real birationally equivalent to a real ruled surface</u>.

<u>Proof</u> : We first note that, since $p_a(C) \leq 0$ we have by the adjunction formula

$$(2.2) \qquad C^2 + C.K \leq -2 \quad \text{or} \quad C^2 + 2 \leq -C.K$$

(K a canonical divisor). By Riemann-Roch (see [Be] I.13), writing $h^0(D)$ for $\dim H^0(X,\mathcal{O}_X(D))$, we have :

$$h^0(C) + h^0(K-C) \geq \chi(\mathcal{O}_X) + 1/2(C^2-C.K) \ .$$

Since X is rational, we have $\chi(\mathcal{O}_X) = 1$ and for the same reason $h^0(K) = 0$. Since C is effective, we also have $h^0(K-C) = 0$. Hence by (2.2), since $C^2 \geq 0$, $h^0(C) \geq C^2 + 2 \geq 2$. In other words, $\dim |C| \geq 1$.

By I.(4.3) we can always choose a real pencil $\{C_\lambda\}$ containing C. Since C is real irreducible, such a pencil has no fixed components. Blowing up the base points of the pencil sufficiently many times we obtain a surface X′, real birationally equivalent to X, on which the strict transforms $\{C'_\lambda\}$ of the C_λ's form a pencil without base points. Clearly, we will again have $p_a(C'_\lambda) \leq 0$.

Since there are no base points, we have $C'^2_\lambda = 0$. In other words the pencil defines a real morphism $X' \longrightarrow \mathbb{P}^1$. By performing a Stein factorization, if necessary, we may assume that the fibres are connected. The generic fibre is real irreducible. If it were of the form D + S(D), then generically a fibre would contain only a finite number of real points and $\dim_K X'(\mathbb{R})$ would be ≤ 1, which is absurd since by construction $\dim_{\mathbb{R}} X'(\mathbb{R}) = \dim_{\mathbb{R}} X(\mathbb{R}) = 2$. Hence the generic fibre is a real irreducible curve. Since $p_a(C'_\lambda) \leq 0$ we must in fact have $p_a(C'_\lambda) = 0$. In other words the generic fibre is smooth and this ends the proof.

We are going to use (2.1) to sharpen V.(1.7).

(2.3) Theorem (Manin) : Let X be a smooth rational real surface, real relatively minimal with $X(\mathbb{R}) \neq \emptyset$. If $K^2 \leqslant 0$ (K a canonical divisor), then X is real birationally equivalent to a real ruled surface.

Proof : Since $K^2 \leqslant 0$, no positive or negative multiple of K can be very ample. In particular, if D is very ample, then $D + nK \neq 0$.

Let D be a very ample S-invariant divisor. For the same reasons as in the proof of V.(1.7), there exists an $n \geqslant 0$ such that the linear system $|D + (n+1)K|$ is empty while $|D + nK|$ is non empty. Since $D + nK \neq 0$, this means that $|D + nK|$ contains an effective S-invariant divisor C (by I.(4.3)).

Since C is effective, $h^0(-C) = 0$. Also, by definition of n, $h^0(C + K) = h^0(D+(n+1)K) = 0$. Applying Riemann-Roch to $-C$, we find :

$$h^0(-C) + h^0(C+K) = 0 \geqslant 1/2(C^2 + C.K) + 1 = p_a(C) .$$

We note as a consequence :

$$(2.4) \qquad C^2 + C.K < 0 .$$

Let $C = \Sigma a_i C_i$ $(a_i > 0)$ be the decomposition of C into a sum of complex irreducible divisors. By the same argument as above applied to $-C_i$ in place of $-C$, we find $p_a(C_i) \leqslant 0$ or, since C_i is irreducible, $p_a(C_i) = 0$. We want to prove that for at least one C_i we have $K.C_i < 0$. Assume that for all i, $K.C_i \geqslant 0$. Then also $K.C \geqslant 0$ and $K.C + C^2 = (n+1)K.C + C.D > 0$ (since D is very ample and C effective). But this contradicts (2.4). Hence the assertion.

By the adjunction formula, this implies $C_i^2 \geqslant -1$.

Now we have two possibilities for C_i :

- either C_i is real, in which case, by the assumption of minimality, we must have $C_i^2 \geqslant 0$ and we are done by (2.1) ;

- or there exists C_j appearing in the decomposition of C such that $C_i \neq C_j$ and $S(C_i) = C_j$. If $C_i^2 = C_j^2 = -1$, then again by the assumption of minimality we must have $(C_i.C_j) > 0$. In all cases we find $(C_i + C_j)^2 \geq 0$. Also since $0 < C_i + C_j \leq C$, the same argument as above shows that $p_a(C_i + C_j) \leq 0$ and we can again conclude by (2.1).

(2.5) Theorem : Let X be as in (2.3). If X is not real birationally ruled then $K^2 > 0$ and $\mathrm{rank}(\mathrm{Pic}(X)^G) = 1$. Moreover $-K$ is ample.

Proof : The condition $K^2 > 0$ is obvious from (2.3).

We note that $K \in \mathrm{Pic}(X)^G$ and that there exists a very ample divisor in $\mathrm{Pic}(X)^G$ (I.(4.4)). If $\mathrm{rank}(\mathrm{Pic}(x)^G) = 1$, then, since $K \neq 0$, nK is very ample for some $n \in \mathbb{Z}$ (recall that since X is rational, Pic(X) is torsion free). Since K is not ample, this means that $-K$ must be ample and proves the last assertion of (2.5).

To prove that $\mathrm{rank}(\mathrm{Pic}(X)^G) = 1$, let D be any S-invariant divisor and H an ample S-invariant divisor. There exists an n such that both nH and D + nH are very ample. In other words, $\mathrm{Pic}(X)^G$ is generated by the classes of ample divisors.

If $\mathrm{rank}(\mathrm{Pic}(X)^G) > 1$, this implies that there exists a very ample divisor which is not of the form mK ($m \in \mathbb{Z}$). We choose such a divisor H. We have $H + mK \neq 0$ for all $m \in \mathbb{Z}$. But now, just as in the proof of (2.3), we can find an n such that $|H + nK|$ contains an effective S-invariant divisor C while $|H + (n+1)K|$ is empty and follow exactly the same proof as in (2.3) to produce a real irreducible divisor satisfying the conditions of (2.1). This proves that we must have $\mathrm{rank}(\mathrm{Pic}(X)^G) = 1$.

(2.6) Corollary : If X is a rational real surface, then X is real birationally equivalent either to :

a) a real ruled surface ;

or to :

b) a so-called real Del Pezzo surface with $\mathrm{rank}(\mathrm{Pic}(X)^G) = 1$ and $-K$ ample.

We will see (see (3.8) below) that, if X is real ruled and real relatively minimal, then $\mathrm{rank}(\mathrm{Pic}(X)^G) = 2$. This means that, unless the real relative-minimal model of X is $\mathbb{P}^2$, the cases a) and b) of (2.8) exclude each other.

3. Birational classification of real ruled rational surfaces

We start by reformulating V.(2.7) in the rational case :

(3.1) Corollary to V.(2.7) (Comessatti) : *Let* $X \longrightarrow \mathbb{P}^1$ *be a real rational ruled surface, then* X *is real birationally equivalent to an affine surface defined in* $\mathbb{R}^3$ *by an equation of the form* :

$$x^2 + y^2 = \pm \prod_{i=1}^{2m} (t-a_i) \qquad a_i \in \mathbb{R}, \text{ distinct}.$$

The classficiation will depend on the m appearing in (3.1). We first note the well known :

(3.2) Proposition : *If* X *is as defined in* (3.1) *with* $m = 1$ *or* $m = 0$ *and* $X(\mathbb{R}) \neq \emptyset$, *then* X *is real birationally equivalent to* $\mathbb{P}^2$.

Proof : If $m = 0$ and $X(\mathbb{R}) \neq \emptyset$ then the ruling has a section and this implies that X is real birationally equivalent to $\mathbb{P}^1 \times \mathbb{P}^1$, hence to $\mathbb{P}^2$. If $m = 1$, then we can find $\alpha \in PGL_1(\mathbb{R})$ such that $\alpha(a_1) = 0$ and $\alpha(a_2) = \infty$. This implies that X is real birationally equivalent to the surface defined by $x^2 + y^2 = t$. But this surface is real birationally equivalent to $\mathbb{P}^2$.

To go further, we will need a fairly obvious lemma that we formulate explicitly.

(3.3) Lemma : *Let* X *be a projective surface and let* X *be real rationally ruled. If the corresponding* m *defined in* (3.1) *is* > 0, *then* $^{\#}X(\mathbb{R}) = m$. *If* $m = 0$, *then* $^{\#}X(\mathbb{R}) = 0$ *or* 1. *In particular, if* $m \geq 2$, *then* m *is a real birational invariant of* X.

We can also reformulate (3.3) in the following way : Let $\pi : X \longrightarrow B$ be a real rational ruled surface. If the number of connected components of $X(\mathbb{R})$, $^{\#}X(\mathbb{R}) = m \geq 2$ and if X_o ($X = X_o \times_{\mathbb{R}} \mathbb{C}$) is B_o-minimal ($B = B_o \times_{\mathbb{R}} \mathbb{C}$) then $2m$ is the number of singular fibres of π.

We also note the following which will be useful :

(3.4) <u>Lemma</u> : <u>The two surfaces defined by</u> :

$$x^2 + y^2 = (t-1)(t+1)(t+½)(t+2)\dots(t+(1/n))(t+n) = f(t)$$

<u>and</u> $x^2 + y^2 = -f(t)$ <u>are real birationally equivalent</u>.

<u>Proof</u> : Explicitly $(x,y,t) \longmapsto (x.t^{-n}, y.t^{-n}, t^{-1})$ defines an isomorphism between the fraction fields of $\mathbb{R}[x,y,t]/x^2+y^2=f(t)$ and $\mathbb{R}[x,y,t]/x^2+y^2=-f(t)$.

(3.5) <u>Theorem</u> : <u>Let</u> $X \longrightarrow \mathbb{P}^1$ <u>and</u> $X' \longrightarrow \mathbb{P}^1$ <u>be two rational ruled surfaces</u>. <u>If for the number of connected components we have</u> $^{\#}X(\mathbb{R}) = {}^{\#}X'(\mathbb{R}) = 2$, <u>then</u> X <u>and</u> X' <u>are real birationally equivalent</u>.

<u>Proof</u> : Let X be as in the theorem and $\pi : X \longrightarrow \mathbb{P}^1$ the ruling. We might as well assume that X is $\mathbb{R}$-relatively minimal (see II.(6.6)). By the above considerations, this means that π has exactly 4 singular fibres over say the points $a_1,\dots,a_4$. Let k be the cross ratio of these 4 points. <u>We will say that</u> k <u>is the cross ratio of the ruling</u>.

We are going to prove that for every couple (k,k') of real numbers there exists a real surface X with $^{\#}X(\mathbb{R}) = 2$ and two rulings $\pi : X \longrightarrow \mathbb{P}^1$ and $\pi' : X \longrightarrow \mathbb{P}^1$, π with cross ratio k and π' with cross ration k'. By (3.1) and (3.4) this will prove the theorem.

We will need :

(3.6) <u>Lemma</u> (Chasles's theorem) : <u>If</u> $x_1,\dots,x_4$ <u>are four points on a</u>

proper conic C and b is any fifth point of C, then the cross ratio of the four lines $x_1b,\dots,x_4b$ is the same for all positions of b on C.

For a proof see Semple and Kneebone [Se&Kn] p. 133.

We can now proceed with the construction of the surface X. For this we fix 4 points in general position (no 3 collinear) in $\mathbb{P}^2(\mathbb{R})$ and a conic C_0 passing through these points. By (3.6), to C_0 and the x_i's corresponds a well defined cross ratio. We choose C_0 so that this cross ratio is equal to k. We also fix for the moment a real point b on C_0 distinct from the x_i's.

Let u be any point in $\mathbb{P}^2(\mathbb{C})$, C_u the conic through u and the x_i's, D_u the line through b and u and let $\{u,T_b(u)\} = C_u \cap D_u$. The correspondence $u \longmapsto T_b(u)$ defines an involution on $U = \mathbb{P}^2(\mathbb{C})\setminus(C_0 \cup_i D_i)$, where D_i is the line through b and x_i.

If j is complex conjugation then $\tau_b = T_b \circ j = j \circ T_b$ is an antiholomorphic involution on U.

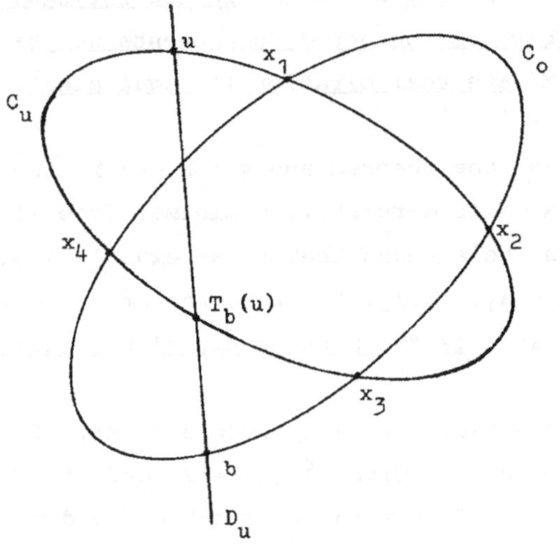

Fig. 5

By construction, T_b and τ_b extend to well defined involutions on the surface X_b obtained by blowing up $\mathbb{P}^2(\mathbb{C})$ in the five points

$b, x_1, \ldots, x_4$. Call S the real structure defined in this way on X_b.

The pencil of lines passing through b defines an S-real one dimensional pencil $\mathfrak{D}$ on X_b without fixed points, hence an S-real ruling of X_b. This pencil has 4 singular fibres $L_i + \tilde{D}_i$ where L_i is the exceptional curve corresponding to x_i and $\tilde{D}_i$ is the strict transform of D_i, the line through x_i and b.

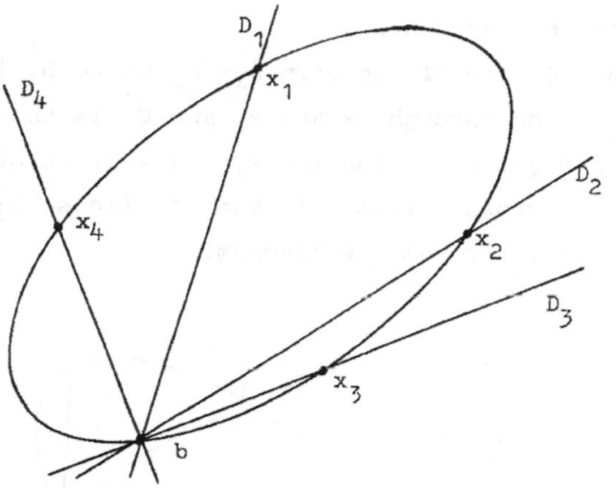

Fig. 6

Since by construction $S(L_i) = \tilde{D}_i$ and $(L_i.\tilde{D}_i) = 1$, the surface X_b is of the desired type.

To compute the cross ratio of the ruling, we note that if B is the exceptional line corresponding to b, then B defines a complex section of the ruling. This means that the cross ratio is given by the four points $\tilde{D}_i \cap B$. But this is nothing but the cross ratio of the four lines D_i, hence k.

Now let C' a second conic passing through the x_i's and corresponding to the cross ratio k'. Changing if necessary, the position of b on C_0 (which does not change the cross ratio k), we can assume that there exists a real tangent to C' through b. Let x be the point of X_b corresponding to the point of contact of this tangent D with C'.

x is S-real so we can blow this point up. Let X be the surface thus obtained.

X inherits from X_b a ruling of cross ratio k. It also has a second ruling defined by the pull back of the pencil of cubics through b, the x_i's and with a double point in x. To see that this ruling is S-real, note that X is isomorphic to a cubic surface in $\mathbb{P}^3$ and that the ruling corresponds to the one induced by the pencil of planes containing the blown up line of x.

This ruling has 5 singular fibres. One is formed of 2 S-real lines (one is the pull back of C′, the other the pull back of D). We do not take this one into account.

The other 4 are of the form $\tilde{D}_i' + \tilde{C}_i'$ where $\tilde{D}_i'$ is the strict transform of the line through x and x_i and $\tilde{C}_i'$ is the strict transform of the conic through x, b and the x_j's $j \neq i$. In other words the cross ratio is the cross ratio of the 4 lines D_i', that is k′ by construction. This proves the theorem.

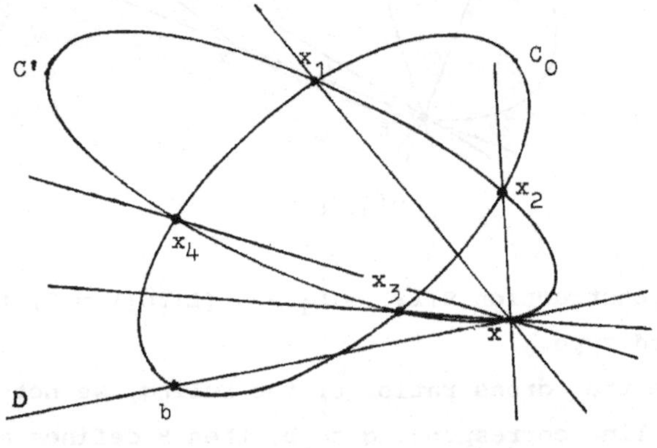

Fig. 7

For $m \geq 3$ the situation is quite different. We do $m \geq 4$ first (for $m = 3$ see (4.7)). We are going to see that in this case there is at most one ruling and obtain a theorem similar to V.(2.8). The results here are due to Comessatti [Co_1] and Iskovskih [Is_1] and [Is_2].

We will first need a detailed study of Pic(X) and of the action of S on Pic(X).

We assume that X in $\mathbb{R}$-relatively minimal but not $\mathbb{C}$-relatively minimal. Let $\{L_i+S(L_i)\}_{1 \leq i \leq 2m}$ be the singular fibres of the ruling. We have $m \geq 1$, since the surface is not $\mathbb{C}$-minimal.

Just as in the irrational case (see proof of V.(3.4)) $Pic(X) = NS(X)$ is generated by the class of a fibre F, the class of a complex section B (this always exists by Tsen's theorem) and the classes of the L_i's. Replacing L_i by $S(L_i) = F-L_i$ if necessary, we may always assume that $S(L_i).B = 1$ and consequently $L_i.B = 0$ (since $B.F = 1$ and $S(L_i).L_i = 1$ -see V.(1.6)).

Now for any divisor D, we must have $D.B = S(D).S(B)$. Writing $S(B) = bB + \Sigma a_i L_i + cF$ and taking D to be successively F, the L_i's and B, it is easy to see using the knowledge of the intersection form ($L_i.S(L_i) = 1$, $L_i^2 = S(L_i)^2 = -1,\ldots$) that we must have :

(3.7) $$S(B) = B - \sum_{i=1}^{2m} L_i + mF \quad .$$

With this description of the action of S on Pic(X), it is easy to see that $Pic(X)^G$ is generated by F and $B + S(B)$, or by F and any element of the form $\pm (2B - \Sigma L_i) + aF$, $a \in \mathbb{Z}$.

As a consequence we have :

(3.8) <u>Lemma</u> (Iskovskih) : <u>Let</u> X <u>be a rational real ruled surface</u>, <u>real relatively minimal but not</u> $\mathbb{C}$-<u>relatively minimal</u>, <u>then</u> $Pic(X)^G$ <u>is generated by the canonical class and the class of a fibre</u>.

<u>Proof</u> : Let X' be the complex surface obtained by blowing down the L_i's . This is a $\mathbb{P}^1$-minimal complex ruled surface with section B', the image of B. This implies (see [Ha] V.2.11 and 2.13) that the cononical class on X' is of the form $K' = -2B' + aF'$ ($a \in \mathbb{Z}$ depending on B^2 - see below). Hence the canonical class on X is of the form $K = -2B + aF + \Sigma L_i$ (see [Ha] V.3.3). By the above, this proves the lemma.

(3.9) <u>Lemma</u> : <u>Let</u> X <u>be a rational real ruled surface with</u> $\#X(\mathbb{R}) = m \geq 2$. <u>Then there exists a complex birational morphism of</u> X <u>onto</u> F_0 <u>(the Hirzebruch surface</u> F_0). <u>Also there exists</u> X' <u>real birationally equivalent to</u> X, <u>such that for all</u> r, $0 \leq r \leq m$ <u>there exists a complex birational morphism from</u> X' <u>onto</u> F_r.

<u>Proof</u> : We may always assume that X is real relatively minimal. Let the L_i's be defined as above and let Y be the complex surface obtained by blowing down the L_i's. This surface is a complex $\mathbb{P}^1$-minimal surface, hence a surface F_n.

Let C be a section in Y such that $C^2 \leqslant 0$. For the strict transform $\tilde{C}$ of C in X, we have again $\tilde{C}^2 = -p \leqslant 0$. Relabeling the L_i's and replacing L_i by $S(L_i)$ if necessary, we may always assume that $L_i.\tilde{C} = 1$ for $1 \leqslant i \leqslant p$ and $L_j.\tilde{C} = 0$ for $p < j \leqslant 2m$. Blowing down the L_i's labeled in this way, we obtain a complex surface Y′ on which the section C′ (the image of $\tilde{C}$) has self-inter-section 0. Hence $Y' = F_0$.

We return to our previous notations. By (3.7), we have $B.S(B) = B^2 + m$. This implies in particular that $B^2 \geqslant -m$. We note that the fibre over a point of intersection of B and S(B) is S-invariant if and only if the point of intersection is real (otherwise we would have $(B+S(B)).F > 2$ or $B.F > 1$, contrary to hypothesis). We can thus perform elementary transformations (see V.(3.5)) centred at the points of intersection of B and S(B) and find X′ real birationally equivalent to X with a section B′ such that $B'^2 = -m$.

Blowing down the $S(L_i)$'s for $1 \leqslant i \leqslant m-r$ and the L_j's for $m-r < j \leqslant 2m$, we obtain a complex $\mathbb{P}^1$-minimal surface containing a section with self-intersection $-r$, hence F_r.

By the proof of (3.9), we can always choose the section B such that $B^2 \leqslant 0$.

(3.10) <u>Corollary</u> : <u>Let</u> X <u>be as in</u> (3.9), <u>real relatively minimal</u>, <u>and let</u> $B^2 = -p \leqslant 0$ (<u>notations as in</u> (3.7)). <u>Then</u> $p \leqslant m$ <u>and</u> :

$$K = -2B - (2+p)F + \sum_{1}^{2m} L_i \ .$$

<u>Moreover</u> $K^2 = 8 - 2m$.

<u>Proof</u> : The inequality $p \leqslant m$ has been proved above. If B is as in (3.10), then blowing down the L_i's we obtain a complex surface F_p. The corollary follows immediately from the classical computation of the

canonical class on such a surface (see [Ha] V.2.11).

(3.11) <u>Lemma</u> : <u>Let</u> X <u>be as in</u> (3.9), <u>real relatively minimal and let</u> C <u>be an</u> S-<u>invariant curve on</u> X <u>and</u> F <u>a fibre</u>. <u>Then there exists</u> X′ <u>real birationally equivalent to</u> X <u>such that if</u> C′ <u>is the proper transform of</u> C <u>and</u> F′ <u>the fibre on</u> X′ <u>then</u> (C′.F′) = (C.F) <u>and</u> C′ <u>has no singular point of multiplicity</u> > 1/2(C′.F′).

<u>Proof</u> : By (3.8) C = aF - bK and since $p_a(F) = 0$, we have by the adjunction formula F.K = -2. Hence C.F = 2b.

Assume C has a singular point x of multiplicity r > b (r ⩽ 2b). If F_x is the fibre through x, C has, appart from x, at most 2b - r < b additional points of intersection with F_x. If x is not real C has also a singularity of multiplicity r at S(x). From the above this implies that x cannot be a non real point of an S-invariant fibre. In other words, it makes sense to speak of an elementary transformation centred at x or of type $elm_{x,S(x)}$ if x ≠ S(x) (see V.(3.5)).

We perform such an elementary transformation. The proper transform C′ of C has, on the corresponding fibre, a singularity of multiplicity strictly less than r and eventually an additional sigularity of multiplicity 2b - r < b. We note that on the surface thus obtained, we have (C′.F′) = (C.F) by construction.

We can repeat this procedure until we obtain a singularity of the desired type and we can do this for every singular point of multiplicity > b. This proves the lemma.

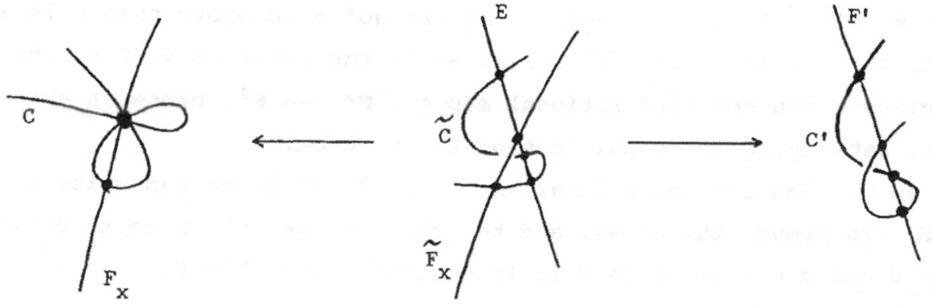

Fig. 8

(3.12) Theorem (Comessatti, Iskovskih) : Let X and X' be two rational real surfaces, real ruled and with for the number of connected components of $X(\mathbb{R})$, ${}^{\#}X(\mathbb{R}) \geq 4$. Then, if $f : X \to X'$ is a real birational equivalence, there exists a real automorphism $g : \mathbb{P}^1 \longrightarrow \mathbb{P}^1$ such that the diagram :

$$\begin{array}{ccc} X & \xrightarrow{\ f\ } & X' \\ \downarrow & & \downarrow \\ \mathbb{P}^1 & \xrightarrow{\ g\ } & \mathbb{P}^1 \end{array}$$

commutes.

Proof : We proceed as in the proof of V.(2.8). By II.(6.4) we can find a smooth real surface X'' and finite sequences σ_1 and σ_2 of monoidal transformations factorizing f

$$\begin{array}{ccccc} & & X'' & & \\ & \swarrow^{\sigma_1} & & \searrow^{\sigma_2} & \\ X & & \overset{f}{\text{-----------}} & & X' \\ \downarrow & & & & \downarrow \\ \mathbb{P}^1 & & & & \mathbb{P}^1 \end{array}$$

Let F' be an S-invariant, irreducible fibre on X', not containing any of the fundamental points of σ_2 (which are in finite number).

Let F'' be the pull back of F' on X''. We have $p_a(F'') = p_a(F') = 0$ and $F''^2 = F'^2 = 0$. Let $C = \sigma_1(F'')$. We are going to prove that C is a fibre of the ruling $X \longrightarrow \mathbb{P}^1$. Just as in the proof of V.(2.8) this will define a non constant rational map $g : \mathbb{P}^1 \longrightarrow \mathbb{P}^1$, hence an automorphism, satisfying the conditions of the theorem.

C is by construction S-invariant on X. By (3.8) we can write $C = aF - bK$. To prove the above, and the theorem, we only need to prove that $b = 0$ and $a = 1$ or since C is irreducible, that $b = 0$.

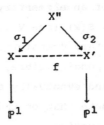

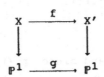

In this direction, we first note that by the same computations as in the proof of (3.11), we have $C.F = 2b$ and this implies $b \geq 0$.

We have, since $F''^2 = 0$ and $p_a(F'') = 0$:

(3.13) (i) $C^2 = \Sigma r_i^2$ and (ii) $p_a(C) = 1/2\, \Sigma r_i(r_i-1)$

where the r_i's correspond to the intersection of F" with the exceptional set of σ_1 in X" (see [Ha] V.3.2, V.3.6 and V.3.7).

If $r_i > 1$ then r_i is the multiplicity of a singular point of C. Hence, by (3.11), we may assume that for all i, $r_i \leq b$. Applying this to (3.13)(i), we get :

$$C^2 \leq b\, \Sigma r_i .$$

On the other hand by (3.13)(ii) and the adjunction formula, we have :

$$C^2 + C.K + 2 = \Sigma r_i(r_i-1) = C^2 - \Sigma r_i$$

or $-C.K \geq \Sigma r_i$ and hence $C^2 + bC.K \leq 0$.

We have also, $C^2 = C(aF - bK) = 2ab - bC.K$ or $C^2 + bC.K = 2ab$. Since $C^2 \geq 0$ this means that $2ab \geq bC.K$. If $b > 0$ we have $0 \geq 2a \geq C.K$.

By direct computation we have :

$$C.K = (aF - bK).K = -2a - bK^2$$

and so : $$0 \geq 2a \geq -bK^2 .$$

But now we recall that by (3.10), $K^2 = 8 - 2m \leq 0$ (since $m = \#X(\mathbb{R}) \geq 4$). If $b > 0$ we obtain a contradiction, hence the only possibility is $b = 0$ and this ends the proof.

(3.14) Corollary 1 : Let X be as in (3.12), then X has, up to an action of $PGL_1(\mathbb{R})$ on the base, at most one ruling.

(3.15) Corollary 2 : Let X and X' be rational real surfaces defined by the affine equations

$$x^2 + y^2 = \prod_{i=1}^{2m} (t-a_i) \text{ and } x^2 + y^2 = \prod_{i=1}^{2m} (t-b_i).$$

If $m \geqslant 4$, then X and X' are real birationally equivalent if and only if there exists $\alpha \in PGL_1(\mathbb{R})$ such that $\alpha(a_i) = b_i$, for some ordering of the a_i's and the b_i's.

4. Real Del Pezzo surfaces.

In relation to (2.6) we will be essentially interested in this part in pinning down real Del Pezzo surfaces (see definition below) for which rank $Pic(X)^G = 1$ and will not attempt to be complete on the subject.

(4.1) Definition : A rational real surface X (see III.(3.1)) is said to be a real Del Pezzo surface if $-K$ (K a canonical divisor) is ample on X.

The number $d = K^2$ is called the degree of the surface.

We have the following characterization of Del Pezzo surfaces (see Manin theorem 24.3 and theorem 24.5 of [Ma_2]) :

(4.2) Theorem : Let X be a Del Pezzo surface over $\mathbb{C}$ of degree d. Then $1 \leqslant d \leqslant 9$. Moreover :

(i) If $d = 9$ then X is isomorphic to $\mathbb{P}^2$;

(ii) If $d = 8$ then X is isomorphic either to $\mathbb{P}^1 \times \mathbb{P}^1$ or to $\mathbb{P}^2$ blown up in one point (that is the Hirzebruch surface F_1) ;

(iii) <u>If</u> $7 \geq d \geq 1$ <u>then</u> X <u>is isomorphic to</u> $\mathbb{P}^2$ <u>blown up in</u> $r = 9-d$ <u>points no three of which lie on a line and no six on one conic</u> (<u>in particular they are all distinct</u>).

The first theorem we want to prove is the following :

(4.3) <u>Theorem</u> (Manin) : <u>Let</u> X <u>be a real Del Pezzo surface of degree</u> d <u>such that</u> $X(\mathbb{R}) \neq \emptyset$. <u>If</u> $d \geq 5$ X <u>is real birationally trivial, that is real birationally equivalent to</u> $\mathbb{P}^2$. <u>If</u> $d \geq 3$, X <u>is real birationally ruled</u>.

To prove (4.3) we will need a lemma. To simplify notations we will call exceptional curve what was called exceptional curve of the first kind.

(4.4) <u>Lemma</u> : <u>If</u> X <u>is a complex Del Pezzo surface of degree</u> $d \geq 3$ <u>then the image in</u> $\mathbb{P}^2$ <u>of an arbitrary exceptional curve on</u> X <u>is of one of the following types</u> :

(i) <u>one of the blown up points</u> x_i ;
(ii) <u>a line passing through two of the points</u> x_i ;
(iii) <u>a conic passing through five of the points</u> x_i.

This is classical for a proof see for example [Ma_2] theorem 26.2 or [Ha] V.4.9.

<u>Proof of</u> (4.3) : We argue for each value of the degree d :

For d = 9 : The result follows from (4.2) and (1.1).

For d = 8 : If X is F_1 then the result follows from (1.2). If X is $\mathbb{P}^1 \times \mathbb{P}^1$ then X is a quadric in $\mathbb{P}^2$ and the result is well known in this case but we will prove it in the next § .

For $d = 7$: The configuration of exceptional curves on X is

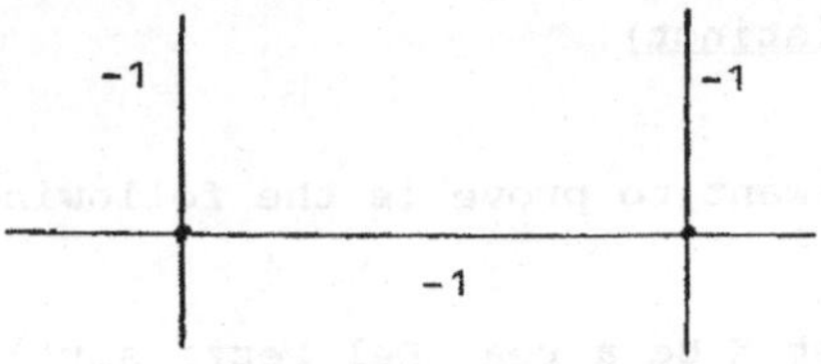

so that the lines on the side are either both fixed or are permuted by the Galois action. In both cases it is possible (by II.(6.2)) to contract them over $\mathbb{R}$ and X is real birationally equivalent to $\mathbb{P}^2$.

For $d = 6$: We may assume that X is real relatively minimal. If not it is real birationally equivalent to a Del Pezzo surface of higher degree for which everything is proved.

We have 6 exceptional lines and they form a hexagone.

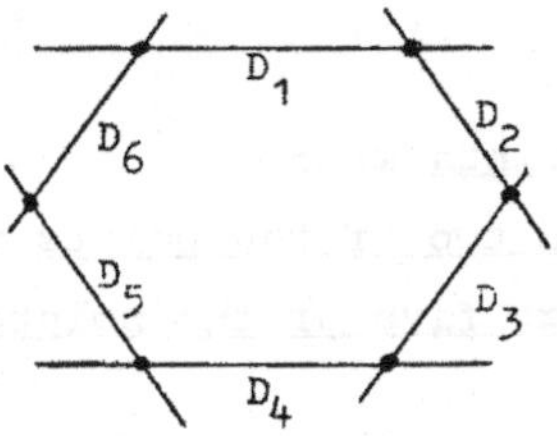

Since X is minimal we must have $D_1 \neq S(D_1)$ and $D_1.S(D_1) > 0$ or $S(D_1) = D_2$ or D_6. Assume $S(D_1) = D_2$. Since $D_2.D_3 = 1$ we must have $S(D_2).S(D_3) = D_1.S(D_3) = 1$. Since $S(D_3) \neq D_2$ this implies $S(D_3) = D_6$. But $D_6.D_6 = 0$ and we can contract these two down over $\mathbb{R}$ contradicting the hypothesis of minimality. In other words X is not real minimal and is real birationally trivial.

For $d = 5$: Since $X(\mathbb{R}) \neq \emptyset$ and X is smooth $X(\mathbb{R})$ is Zariski dense in X. In particular there exists a point $x \in X(\mathbb{R})$ not contained in any of the exceptional curves. We blow up this point and obtain a real surface $(\tilde{X},\tilde{S})$ with an $\tilde{S}$-real exceptional line E (image of x). By (4.4) this line intersects 5 other exceptional lines, corresponding to the 4 lines in $\mathbb{P}^2$ passing through the image of x in $\mathbb{P}^2$ and the points $x_1,\ldots,x_4$ (where X is $\mathbb{P}^2$ blown up in the x_i's), and intersects the conic through the image of x and the x_i's. Since E is real these 5 lines are globally $\tilde{S}$-invariant and do not intersect. By II.(6.2) we can contract all 5 over $\mathbb{R}$ and obtain $\mathbb{P}^2$. This proves the theorem in this case.

We have of course proved that for $d \geqslant 5$, X is real birationally ruled. For d = 3 or 4 we only need, by (2.6), to prove that rank $\mathrm{Pic}(X)^G > 1$.

Let X be, over $\mathbb{C}$, $\mathbb{P}^2$ blown up in the x_i's (5 or 6 of these). Pic(X) is then generated by H, the pull back of a general line in $\mathbb{P}^2$ and the L_i's (where L_i is the exceptional curve corresponding to x_i). For the real structure S on X, $S(L_i)$ must also be an exceptional curve on X. If we write in terms of the class in Pic(X) the only possibilities are, by (4.4) : L_i, another L_j, $H-L_k$ for some k or $2H - \Sigma L_i$. But the canonical class is $-3H + \Sigma L_i$ and it is easily checked that $L_i + S(L_i)$ and K are linearly independant in Pic(X). This implies in particular that rank $\mathrm{Pic}(X)^G > 1$ and ends the proof of (4.3).

(4.5) <u>Corollary</u> : <u>Let</u> X <u>be a rational real surface real relatively minimal with</u> $X(\mathbb{R}) \neq \emptyset$. <u>If</u> X <u>is not real birationally ruled, then</u> X <u>is a Del Pezzo surface of degree</u> 1 <u>or</u> 2.

We are now going to prove that such surfaces actually exist by constructing explicit examples.

Let $x_1,\ldots,x_7$ be 7 real points in general position in $\mathbb{P}^2(\mathbb{R})$ (that is in this case, that satisfy the conditions of (4.2)(iii), see Demazure [Dem] III) and let x be any point in $\mathbb{P}^2(\mathbb{C})$.

The linear system of cubics C_x , passing through the x_i's and x has dimension 1 (i.e. is a pencil) and has an unassigned base point (see [Ha] V.4.5). In other words there is a uniquely determined ninth point $T(x)$ such that for any two cubics C and C' in C_x, $C \cap C' = \{x_1,\dots,x_7,x,T(x)\}$.

Let C_i be the cubic through the x_j's and with a double point at x_i. The correspondence $x \longmapsto T(x)$ (which is defined everywhere except at the x_i's) defines an involution on $\mathbb{P}^2 \setminus \bigcup_i C_i$ (the Geiser involution). This involution extends to a well defined involution on the surface X obtained by blowing up $\mathbb{P}^2$ in the 7 points x_i. (In fact there is another way of defining this involution. Consider the morphism φ_{-K} associated to $-K$, then it is possible to prove that $\varphi_{-K} : X \longrightarrow \mathbb{P}^2$ realises X as a double covering of $\mathbb{P}^2$. The involution associated to this double covering is the Geiser involution see [Dem] V §4).

Call $\tau = T \circ j = j \circ T$ (j complex conjugation). Then τ also extends to a well defined anti-holomorphic involution on X and hence defines a real structure S on X.

We are going to prove that for this real structure we have $\mathrm{rank}(\mathrm{Pic}(X)^G) = 1$.

For this we recall that $\mathrm{Pic}(X)$ is again generated by the class H of the pull back of a general line in $\mathbb{P}^2$ and the class L_i of the exceptional lines corresponding to the x_i's. The canonical class is $K = -3H + \Sigma L_i$.

Let C_i be as above, the cubic in $\mathbb{P}^2$ passing through the x_j's and with a double point in x_i. The class in $\mathrm{Pic}(X)$ of the strict transform of C_i is $3H - 2L_i - \sum_{i \neq j} L_j$. By the description of S given above, this is $S(L_i)$. In other words we have : $L_i + S(L_i) = -K$.

To end the proof we must compute $S(H)$. For this let D be the line in $\mathbb{P}^2$ passing through x_1 and x_2. If x lies on D (distinct from x_1 and x_2) then by the generalization of Pascal's theorem $T(x)$ lies on the conic through the 5 remaining points and conversely. This means that we have :

$$S(H-L_1-L_2) = 2H - \sum_{i \geqslant 3} L_i$$

or

$$S(H) = -2K + 2H - \Sigma L_i = 8H - 3\,\Sigma L_i\ .$$

In other words, we have $H + S(H) = -3K$.

From here it is easy to prove that $\mathrm{Pic}(X)^G$ is generated by K (or -K). In fact we can be even more precise and compute also the Comessatti characteristic λ (see I.§3) which is equal to 1 here. To see this consider the basis $\{-K, H+K, H+K-L_1, \ldots, H+K-L_6\}$ of Pic(X) and write down the matrix of S in this basis.

We have constructed above an example for $d = 2$. We now proceed to build one for $d = 1$.

For this consider 8 real points $x_1, \ldots, x_8$ in general position in $\mathbb{P}^2(\mathbb{R})$ (here for general position we must add to the conditions of (4.2)(iii) the condition that the 8 points do not lie on a cubic with a double point in one of them).

We are going to consider Bertini's involution defined as follows : Let x be any point in $\mathbb{P}^2$ distinct from the x_i's and let $\mathfrak{S}_x$ be the linear system of sextics passing through x and having a double point at each of the x_i's. Let C be the cubic passing through x and the x_i's, then there is a well defined point T(x) such that for any $B \in \mathfrak{S}_x$, $B \cap C = \{2x_1, \ldots, 2x_8, x, T(x)\}$ (see Dieudonné [Di] VI.57). In the same way as in the preceding case this defines a real structure S on the surface X obtained by blowing up the x_i's. Here φ_{-2K} realises X as a double covering of a quadric cone and defines the Bertini involution on X (see [Dem] V.§5).

Using of the same notations as before we have, $S(L_i)$ is the pull back of the sextic with a triple point in x_i and double points in the other x_j's. Hence $S(L_i) = 6H - 3L_i - 2 \sum_{i \neq j} L_j$. Recalling that for the canonical divisor we have $K = -3H + \Sigma L_i$, we find $L_i + S(L_i) = -2K$.

To compute S(H) we use the fact that $3H - \Sigma L_i = -K$ and $S(K) = K$. From this and the computation of $S(L_i)$ we get :

$$3S(H) = -17K - \Sigma L_i \text{ or } 3H + 3S(H) = -18K$$

hence $H + S(H) = -6K$.

Again it is easy from here to prove that $\mathrm{Pic}(X)^G$ is generated by K. The Comessatti characteristic is this time equal to zero (to see this use the basis $(-K, H+3K, L_1+K, \ldots, L_7+K))$.

We have not been able to prove that real Del Pezzo surfaces with $\mathrm{rank}(\mathrm{Pic}(X)^G) = 1$ where always of this type but we can prove that the invariants computed above are general :

(4.6) <u>Theorem</u> (Comessatti) : <u>Let</u> X <u>be a rational real surface</u>, <u>real relatively minimal with</u> $X(\mathbb{R}) \neq \emptyset$ <u>and not real birationally ruled</u>. <u>Then</u> X <u>is by</u> (4.5) <u>a Del Pezzo surface of degree</u> d= 1 <u>or</u> 2.

<u>If</u> d = 2 <u>then we have</u> $B_2 = 8$, $b_2 = 7$ (<u>see</u> I.(2.1)) <u>and</u> $\lambda_2 = 1$ (<u>see</u> I.(3.4)). <u>In particular</u> $X(\mathbb{R})$ <u>has</u> 4 <u>components</u>.

<u>If</u> d = 1 <u>then we have</u> $B_2 = 9$, $b_2 = 8$ <u>and</u> $\lambda_2 = 0$. <u>In particular</u> $X(\mathbb{R})$ <u>has</u> 5 <u>components</u>.

<u>Proof</u> : We recall (see III.(4.1.1.), that we have an isomorphism of Galois structures :

$$\mathrm{Pic}(X) = H^2(X(\mathbb{C}),\mathbb{Z})(1).$$

Recall also that $\mathrm{rank}(\mathrm{Pic}(X)) = 8$ if d = 2 or 9 if d = 1. This proves the assertion about B_2.

Since $\mathrm{rank}(\mathrm{Pic}(X)^G) = 1$ we have $b_2 = 8-1 = 7$ if d = 2 or $9-1 = 8$ if d = 1 (recall I.(4.8)). This proves the assertion about b_2.

For λ_2 we note that we must have by definition $\lambda_2 \leqslant \mathrm{Inf}(b_2, B_2-b_2) = 1$. Hence $\lambda_2 = 0$ or 1. But by II.(2.6) and III.(3.2) we have $\lambda_2 \equiv b_2$ (mod.2). This proves the assertion about λ_2.

By I.(3.7) $\dim H^1(G,\mathrm{Pic}(X)) = \dim H^2(G,H^2(X(\mathbb{C}),\mathbb{Z})) = b_2 - \lambda_2$. The above imply in this case that

$$\dim H^1(G,\mathrm{Pic}(X)) = 6 \text{ if } d = 2 \text{ or } 8 \text{ if } d = 1.$$

The assertion about the number of connected components follows from III.(3.3).

Now we have introduced the Geiser involution we can prove

(4.7) Proposition : Let X and X' be as in (3.15) but with $m = 3$ ($^{\#}X(\mathbb{R}) = {}^{\#}X'(\mathbb{R}) = 3$) . Then X and X' are $\mathbb{R}$-birationally equivalent if and only if there exists $\alpha \in PGL_1(\mathbb{R})$ such that $\alpha(\{a_i\}) = \{b_i\}$.

For the proof we will need a lemma

(4.8) Lemma (Manin-Iskovskih) : Let X and X' be two Del-Pezzo surfaces of degree $d \leq 4$ and let $f : X \to X'$ be a birational equivalence, then f decomposes into

$$X \longrightarrow \dots \longrightarrow X_{i-1} \xrightarrow{f_i} X_i \longrightarrow \dots \longrightarrow X'$$

where f_i is either an isomorphism (biregular) or X_{i-1} is ruled and f_i is an elementary transformation in the ruling (that is, not centred at a point in a singular fibre).

See Colliot-Thélène [Col].

Proof of (4.7) : Let again B be a complex section of the fibration and $L_i + S(L_i)$ the singular fibres. By the proof of (3.9) we can always assume that $B^2 = -1$ and $B.L_i = 0$ for all i. In this case X is a complex Del Pezzo surface of degree 2.

By (4.8) we only need to prove that if $\pi' : X \to \mathbb{P}^1$ is a second ruling of X then there exists an automorphism g of X such that $\pi' = \pi \circ g$. In fact we are going to prove that there exists exactly 2 rulings and that they are exchanged by the Geiser involution.

For this we contract over $\mathbb{C}$, B and the L_i's. We obtain in this way $\mathbb{P}^2$. Let H be the pull back in X of a general line in $\mathbb{P}^2$. The general fibre F of π has class $H-B$. With this in mind we can give the classes of the 56 exceptional curves in X. These are : B, the 6 L_i's, 6 $D_i = F-L_i$, 15 $D_{ij} = F+B-L_i-L_j$, 6 $C_i = 2F+2B-\sum_{k\neq i} L_k$, 15 $C_{ij} = 2F+B-\sum_{k\neq i,j} L_k$, 6 $\Gamma_i = 3F+2B-\sum_{k\neq i} L_k-2L_i$ and $\Gamma = 3F+B-\Sigma L_i$.

Recalling that $S(F) = F$, $S(L_i) = F-L_i$ and the value of $S(B)$ given in (3.7), it is easy to see that, $S(B) = \Gamma$, $S(L_i) = D_i$, $S(D_{ij}) = C_{ij}$ and $S(C_i) = \Gamma_i$.

Now we know that if there exists a second fibration it must have 6 singular fibres (by (3.3)) and that by V.(1.6) the fibre must be of the form $E+S(E)$ with E exceptional and $(E.S(E)) = 1$.

Checking with the above it is easy to see that the only possibilities are $L_i+ D_i$ and $C_i+ \Gamma_i$.

But the explicit description of the action of the Geiser involution on exceptional curves is $E \mapsto -K-E$ (see [Ma_1] II § 4.7). Since $K = -3F-2B+\Sigma L_i$ it is easy to see that the assertion is proved and hence the proposition.

5. <u>Real quadric and cubic surfaces</u>

The results in this part are all completely classical and are only included here as illustrations of the methods developed in the preceding parts. We will practically give no proofs and most of the results should be considered as exercices.

The classification of projective real quadric surfaces is equivalent to the classification of real quadratic forms in 4 variables with the obvious restriction that the forms $q(x,y,z,w)$ and $-q(x,y,z,w)$ define the same surface.

Since we are only interested here in smooth surfaces the corresponding quadratic forms are non-degenerate (that is the associated bilinear form is of maximal rank). From this it follows that we have, up to real isomorphy (and even projective equivalence) exactly 3 real smooth quadrics in $\mathbb{P}^3$.

(5.1.1.) The surface defined by $x^2+y^2+z^2+w^2 = 0$. In this case we have $X(\mathbb{R}) = \emptyset$. Over $\mathbb{C}$ this surface is $\mathbb{P}^1 \times \mathbb{P}^1$ and $\mathrm{Pic}(X)$ is generated by the two factors. As usual we write (a,b) for a divisor of class $a(\mathbb{P}^1\times \{0\}) + b(\{0\} \times \mathbb{P}^1)$. A line of class $(1,0)$ (resp. $(0,1)$ is transformed by S into a distinct line of type (1,0) (resp. (0,1)). In particular the Galois action on Pic(X) is trivial.

(5.1.2.) The surface defined by $x^2+y^2+z^2-w^2 = 0$. The real part $X(\mathbb{R})$ of this surface is S^2 (= T_0). Taking the affine equation $x^2+y^2 = (w-1)(w+1)$ we have already studied this type of surface in §3. To have another description we take the generators of Pic(X) to be the lines $A = \{(x=iy) \cap (z=w)\}$ and $B = \{(x=-iy) \cap (z=w)\}$. Obviously $S(A) = B$ and this describes the Galois action on Pic(X).

(5.1.3) The surface defined by $x^2+y^2-z^2-w^2 = 0$. This surface is isomorphic over $\mathbb{R}$ to $\mathbb{P}^1 \times \mathbb{P}^1$ and the real part is a torus T_1. In this case, just as in (5.1.1) the Galois action on Pic(X) is trivial.

We conclude on quadrics with :

(5.2) <u>Harnack's theorem for curves on a quadric</u> (Hilbert) : <u>Let</u> C <u>be a smooth real curve of degree</u> d <u>on a real quadric surface</u>. <u>Then</u> $C(\mathbb{R})$ <u>has at most</u> $d^2/4 - d + 2$ <u>components if</u> d <u>is even or at most</u> $(d^2-1)/4 - d + 2$ <u>if</u> d <u>is odd</u>.

<u>Proof</u> : Let the curve C have type (a,b) (see above or see [Ha] III.Ex.5.6). We have $d = a+b$ and for the genus $g(C) = ab-a-b+1$. Recalling from Harnack's theorem that $^{\#}C(\mathbb{R}) \leqslant g(C)+1$ we get the result.

For cubic surfaces we recall that over $\mathbb{C}$ they are characterized as being $\mathbb{P}^2$ blown up in 6 points in general position. In particular a cubic surface has 27 lines. These are exceptional and are the only exceptional curves on the surface. If such a surface has a real structure (X,S), then 27 being odd, one of the exceptional lines is S-invariant. Hence we can blow down one of these over $\mathbb{R}$. In other words as far as birational classification is concerned, cubic surfaces have the same birational types as Del Pezzo surfaces of degree 4.

By (4.3), (3.5) and (3.2) this means :

(5.3) <u>Proposition</u> : <u>Real cubic surfaces fall into two birational</u>

<u>types</u>. <u>One is trivial</u> (i.e. $\mathbb{P}^2_{\mathbb{R}}$) <u>the other is the type defined in</u> (3.5).

From (5.3) it is easy to give a complete description of all real cubic surfaces.

(5.4.1.) $\mathbb{P}^2$ blown up in 6 real points. By II.(6.9) $X(\mathbb{R}) \cong U_6$ (see II.(6.7). The 27 lines on X are real and the Galois action on Pic(X) is trivial.

(5.4.2) $\mathbb{P}^2$ blown up in 4 real points and 2 complex conjugate. $X(\mathbb{R}) \cong U_4$ and X has 15 real lines (use (4.4)). The invariants for the action of S on Pic(X) (see I.(3.5) and (3.6)) are $b = \mathrm{rank}(\mathrm{Pic}(X)^G) = 6$ and $\lambda = 1$ (to see this one must take into account the contribution of the sum of the two lines corresponding to the complex points).

(5.4.3) $\mathbb{P}^2$ blown up in 2 real points and 4 complex, two by two conjugate. We have $X(\mathbb{R}) \cong U_2$ and X has 7 real lines. The invariants for the action of S on Pic(X) are $b = 5$ and $\lambda = 2$.

(5.4.4) $\mathbb{P}^2$ blown up in 6 complex points two by two conjugate. We have $X(\mathbb{R}) \cong U_0$ ($\cong \mathbb{R}^2(\mathbb{R})$ (Caution this is $\mathbb{P}^2(\mathbb{R})$ topologically but not geometrically). There are 3 real lines (corresponding to the pull back of the lines through the conjugate paires). The invariants for the action of S on Pic(X) are $b = 4$ and $\lambda = 3$.

(5.4.5) A surface obtained by blowing up a real point (not on an exceptional curve) of a surface of the type constructed in the proof of (3.5). We have $X(\mathbb{R}) \cong T_0 \coprod \mathbb{P}^2(\mathbb{R})$ (to see this use II.(6.9) and (6.11) or see next §.). There are 3 real lines on X coming respectively from the exceptional line associated to the blown up point x and using the notations of the proof of (3.5) the line through b and x (where we have again written x for its image in $\mathbb{P}^2$) and the conic through $x_1,\ldots,x_4$ and x.

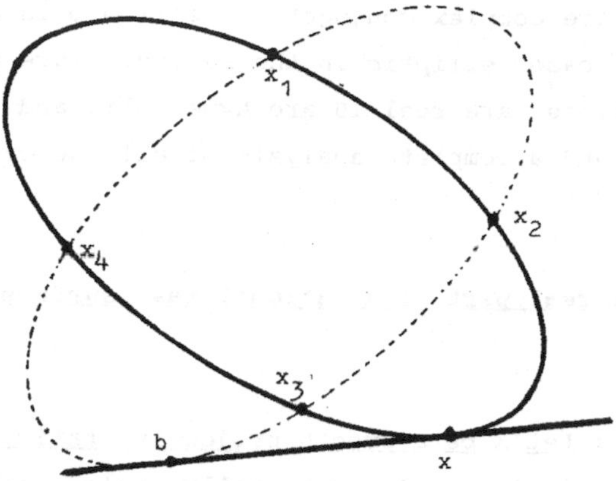

Figure 11

The invariants for the action of S on Pic(X) are b = 3 (to see this use (3.8)) and $\lambda = 2$.

We note that from the preceding description follows a fact quoted earlier :

(5.5) Let X be a real cubic surface then $\text{rank}(\text{Pic}(X)^G) \geqslant 3$.

Much more can be said about real cubic surfaces and for further results we refere to the book of Segre [Seg]. One interesting point to be noted and that is rarely mentioned is the following : Contrarely to the complex case where the 27 lines are undistinguishable from one another, the real lines even when all the 27 are real are not all equivalent. We can describe the difference in the following way : Let D be a line in X and P a plane in $\mathbb{P}^3$ containing D. Then $P \cap X = \{D \cup C\}$ where C is a conic. Since X is non singular, through each point of D passes one and only one such conic (defined by the tangent plane to X at that point). This pencil of conic's defines an involution on D. We have two possibilities either the involution has real fixed points (which correspond to a conic tangent to D, eventually degenerate) or the fixed points are complex conjugate. The line D is called hyperbolic in the first case elliptic in the second. Segre proves in [Seg] that if the 27 lines are real 15 are hyperbolic and 12 are elliptic (for more details and a complete analysis of all the cases see [Seg]).

6. Topology of the real part of a rational real surface II.

(6.1) Proposition : Let X be a smooth projective rational real surface, real relatively minimal, real birationally ruled with $^{\#}X(\mathbb{R}) = m \geqslant 2$ components. Then $X(\mathbb{R})$ is the disjoint union of m spheres and $m = \frac{1}{2}\mathrm{rank}(\mathrm{Pic}(X)) - 1 = \frac{1}{2}\mathrm{rank}(NS(X))-1$.

Proof : We recall ((3.8)) that $\mathrm{Pic}(X)^G$ is generated by K and F. Since $K^2 = 8-2m$ (3.10) and $F^2 = 0$, $X(\mathbb{R})$ is orientable by III.(4.2). Also since $F = L_1 + S(L_1)$ and $-K = B + S(B) + aF$ (see (3.7) and (3.10)) we have for the Comessatti characteristic $\lambda = 2$. Hence $H^2(G,\mathrm{Pic}(X)) = 0$. By III.(4.1) this means that $H^1(X(\mathbb{R}),\mathbb{Z}/2) = 0$. In other words the only possible components are spheres.

The last assertion follows from I.(3.7) and III.(3.3).

(6.2) Corollary : Let $\mathcal{Y}_1,\ldots,\mathcal{Y}_m$ be an arbitrary family of non-orientable topological surfaces (compact, connected and without boundary). Then there exists a rational real surface such that :

$$X(\mathbb{R}) = \mathcal{Y}_1 \amalg \ldots \amalg \mathcal{Y}_m .$$

Proof : Immediate by (6.1) and II.(6.9).

(6.3) Proposition : Let X be a smooth projective rational real surface, real relatively minimal and such that X is not real birationally ruled. Then $X(\mathbb{R}) \cong 4T_0$ ($\cong 4S^2$) or $X(\mathbb{R}) \cong 4T_0 \amalg \mathbb{P}^2(\mathbb{R})$.

Proof : The invariants for the Galois action on Pic(X) were computed in (4.6). They are $\mathrm{rank}(\mathrm{Pic}(X)) = 8$, $\mathrm{rank}(\mathrm{Pic}(X)^G) = 1$ and $\lambda = 1$ if X is a Del Pezzo surface of degree 2. By III.(3.3) we find $^{\#}X(\mathbb{R}) = 4$ in this case and by III. (4.1.1) and I.(3.7), $H^1(X(\mathbb{R}),\mathbb{Z}/2) = H^2(G,\mathrm{Pic}(X)) = 0$.

If X is a Del Pezzo surface of degree 1, then $\mathrm{rank}(\mathrm{Pic}(X)) = 9$, $\mathrm{rank}(\mathrm{Pic}(X)^G) = 1$ and $\lambda = 0$. By the same argument as above $^{\#}X(\mathbb{R}) = 5$ and $H^1(X(\mathbb{R}),\mathbb{Z}/2) = H^2(G,\mathrm{Pic}(X)) = \mathbb{Z}/2$. The only possibilities are the ones given in (6.3).

(6.4) Proposition : *Let X be a smooth projective rational real surface, real relatively minimal such that $^{\#}X(\mathbb{R}) = 1$. Then X is real birationally trivial (real birtionally equivalent to $\mathbb{P}^2$) and $X(\mathbb{R}) = U_0$ (projective plane) or T_1 (a torus), or U_1 (a Klein bottle) or T_0 (a sphere).*

Proof : The first assertion follows from (3.2), (4.5) and (4.6) (or (6.3)). The second from (1.1), (1.3), (5.1.2) and (5.1.3).

(6.5) Corollary (Colliot-Thélène's conjecture) : *Let X be a smooth projective real rational surface. Then X is real birationally trivial if and only if X(ℝ) is connected.*

Proof : The "only if" follows from II.(6.12) and the "if" from (6.4).

Bibliographical Notes :

§1 comes from Manin [Ma_1]. All the results in §2 except, maybe, the explicit formulation of (2.6) also comes from Manin [Ma_1].

§3 with the exeption of (3.5) (which strangely seems either to have been unknown or, more likely, to have been forgoten) comes from Comessatti [Co_1] and Iskovskih [Is_1] and [Is_2].

Everything in §4 comes from Manin [Ma_1] except for (4.6) which can be found in Comessatti, although the proof given by Comessatti is considered by many as undecipherable (see for example Iskovskih [Is_2]). It is true that [Co_1] contains some very ambiguous statements (see below). Nevertheless (4.6) is explicitly in [Co_1], except for the computation of the invariants for the action of S on Pic(X)). In fact there is a discrepancy between our computations and the computations made by Comessatti. The difference comes from the fact that Comessatti does not consider the real minimal model (contrary to what he seems to claim ? there is a hint that he does, since he claims that the Del Pezzo surface of degree one he considers does not have an elliptic fibration) but the model obtained by blowing up the base points of the linear system of cubics in case d = 2 (resp. of the linear system of sextics in case d = 1) (see the explicit examples constructed).

Results on Quadrics are very old and can be traced back to the XVIIIth century. Results on cubics date from the XIXth century. The most complete reference remains B. Segre [Seg] (see also Manin [Ma_2]).

(6.5) was conjectured by Colliot-Thélène, private communication.

VII. REAL ELLIPTIC SURFACES

1. Local theory of real elliptic fibrations with a section.

Let X_o be a smooth surface over $\mathbb{R}$ (see remark on notations I.(1.15)) and $\pi : X_o \longrightarrow B_o$ a real elliptic fibration of X_o over a curve B_o. Replacing eventually π by a Stein factorization we may assume that $\pi_*(\mathcal{O}_{X_o}) = \mathcal{O}_{B_o}$. In what follows we will always make this assumption.

Our aim here is to classify the fibres $\pi^{-1}(b)$ when $b \in B_o(\mathbb{R})$ and to study how $p^{-1}(b)$ varies with b.

We will first examine the case when the fibration has a section $s : B_o \longrightarrow X_o$.

We need to introduce some notations : Let η be a generic point of B_o and X_η the corresponding generic fibre $\pi^{-1}(\eta)$. By definition X_η is a curve of genus 1 over the residue field $\kappa(\eta)$ (or $\mathbb{R}(\eta)$).

We consider $\mathrm{Pic}^o_{X_\eta|\kappa(\eta)} = \mathrm{Jac}(X_\eta)$. By the general theory of minimal models of curves (see, for example, since here, we are in genus one, Neron [Né]). We can associate to Jac (X_η) a minimal model $\tilde{X}_o$ over B_o. By construction we can again view $\tilde{X}_o$ as an elliptic surface over $\mathbb{R}$ fibred over B_o with fibration $\tilde{\pi} : \tilde{X}_o \longrightarrow B_o$. We will refer to this fibration as the <u>jacobian fibration</u> of $\pi : X_o \longrightarrow B_o$. By definition of Jac, this fibration has a section $s : B_o \longrightarrow \tilde{X}_o$ and it is unique up to a B_o-isomorphism (see V.(1.4)). Moreover if $\pi : X_o \longrightarrow B_o$ has a section and X_o is non singular and B_o-minimal then π is isomorphic to its jacobian fibration (see [Né] or for a proof in the complex case [B&P&V] p. 154).

Call $B_o^\#$ the set of points in B_o over which $\pi^{-1}(b)$ is non singular. We have :

(1.1) <u>Lemma</u> : <u>Let</u> $\pi : X_o \longrightarrow B_o$ <u>be a real elliptic fibration of a</u>

smooth surface X_o *(defined over* $\mathbb{R}$*), satisfying* $\pi_* \mathcal{O}_{X_o} = \mathcal{O}_{B_o}$. *Let* $\tilde{\pi} : \tilde{X}_o \longrightarrow B_o$ *be the corresponding jacobian fibration. Then if* $b \in B_o^{\#} \cap B(\mathbb{R})$ *we have :*

$$\pi^{-1}(b) = X_b = \mathrm{Pic}^o_{X_b|\kappa(b)} = \mathrm{Jac}(X_b).$$

Proof : This is in fact, nothing but a special and well known case of a far more general theorem of Raynaud ([Ra] theorem 8.2 p. 66, see Deligne-Mumford [De&Mu]).

We will also need the following : Let X be a complex surface and $\pi : X \longrightarrow B$ an elliptic fibration. We can also in this case, define the jacobian fibration $\tilde{\pi} : \tilde{X} \longrightarrow B$. Then, if $\pi^{-1}(b)$ is not a multiple fibre, $\tilde{\pi}^{-1}(b)$ is isomorphic to $\pi^{-1}(b)$ and if $\pi^{-1}(b) = \mu\Gamma$, then Γ is a quotient of $\tilde{\pi}^{-1}(b)$ (see [Sha] Chap. VII, corollary 1 and 2 of theorem 7). This of course applies to the case $X = X_o \times_{\mathbb{R}} \mathbb{C}$ where the jacobian fibration of X is obtained from the jacobian fibration of X_o by extending scalars to $\mathbb{C}$.

A very simple remark which will be useful in the sequel is :

(1.2) *Lemma* : *Let* E *be a curve of genus* 1 *over* $\mathbb{R}$ *such that* $E(\mathbb{R}) = \emptyset$, *then* $\mathrm{Jac}(E)(\mathbb{R})$ *has two components.*

Proof : This is well known and comes from the fact that the Comessatti characteristic λ_1 for the action of S on $H^1(E(\mathbb{C}),\mathbb{Z})$ is zero in this case (see [Si_1] proposition 10 or for a different approach Knebusch [Kn]). The result follows from IV.(1.9).

Let $\pi : X_o \longrightarrow B_o$ be a real elliptic fibration with a (real) section. This implies that the generic fibre X_η considered as a curve over $\kappa(\eta) = \mathbb{R}(B_o)$ has an $\mathbb{R}(B_o)$-rational point. From the classical theory of elliptic curves this means that the equation of X_η over $\mathbb{R}(B_o)$ can be normalised in Weierstrass form :

$$y^2z = x^3 + p(b)xz^2 + q(b)z^3$$

with p and q in $\mathbb{R}(B_o)$.

We will denote $\Delta = 4p^3+27q^2 \in \mathbb{R}(B_o)$.

(1.3) Lemma : Let $y^2 = x^3 + ax + b$ $(a,b \in \mathbb{R})$ be the Weierstrass equation of an elliptic curve E over $\mathbb{R}$. Let $\Delta = 4a^3 + 27b^2$. Then the sign of Δ is an invariant of E over $\mathbb{R}$ and if $\Delta < 0$, $E(\mathbb{R})$ has two connected components, if $\Delta > 0$, $E(\mathbb{R})$ has only one.

Proof : The number of components being a real invariant of E it is sufficient to prove the last assertion. This in turn follows from the fact that the number of components only depends on the number of real zeros of the equation $x^3 + ax + b = 0$ and on the fact that Δ is the discriminant of the equation.

Let $b_0 \in B_o(\mathbb{R})$ and let v be the valuation of $\mathbb{R}(B_o)$ at b_0. We have :

(1.4) Corollary : The number of connected components of $(\pi^{-1}(b))(\mathbb{R})$ $(b \in B_o(\mathbb{R}))$ changes in the neighbourhood of b_0 in $B_o(\mathbb{R})$ if and only if Δ changes signs at b_0. In other words if and only if $v(\Delta) \equiv 1 \pmod{2}$.

Throughout the remainder of this section we assume that $\pi : X_o \longrightarrow B_o$ is a real elliptic fibration with a section $s : B_o \longrightarrow X_o$. We will say that $b_0 \in B_o$ is a special point if the fibre over b_0 is singular.

Let t be a local parameter of B_o at $b_0 \in B_o(\mathbb{R})$. Then t defines a local parameter of $B = B_o \times_{\mathbb{R}} \mathbb{C}$ at b_0 and also, since b_0 is a smooth point, a local real analytic parameter of $B(\mathbb{R})$. Let D be a small disc on $B(\mathbb{C})$, associated to t and such that D contains no special points outside b_0. The equation of $\pi^{-1}(D\backslash\{b_0\})$ in $\mathbb{P}^2 \times (D\backslash\{b_0\})$ can be written in the form :

$$y^2 z = x^3 + p(t)xz^2 + q(t)z^3$$

with p and q real functions of t.

We are now ready to state and prove our main theorem concerning

the structure of the fibres of an elliptic surface :

(1.5) <u>**Theorem**</u> **:** <u>**Let**</u> $\pi : X_o \longrightarrow B_o$ <u>**be a**</u> <u>B_o**-minimal real elliptic fibration with a section**</u> $s : B_o \longrightarrow X_o$, X_o <u>**smooth and**</u> $B_o(\mathbb{R}) \neq \emptyset$. <u>**Let**</u> b_0 <u>**be a special point of**</u> $B_o(\mathbb{R})$ <u>**and let notations be as above**</u>. <u>**Then the behaviour of**</u> X_b <u>**the fibre over**</u> b <u>**when**</u> $b \in U = D \cap B_o(\mathbb{R})$, <u>**and the geometric structure of**</u> $X_{b_o} \times_{\mathbb{R}} \mathbb{C}$ <u>**are described in the following table :**</u>

Néron's classification	Kodaira's notations	The number of components of X_b, $b \in U$: does not change and is equal to	changes	Geometric structure of X_{b_o}	$\chi(X_{b_o}(\mathbb{R}))$
$b_m \begin{cases} v(j)=-m<0 \\ v(q) \text{ even} \end{cases}$ m even	I_m m even	2		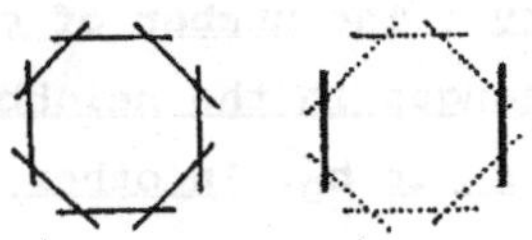 (m components)	-m or 0
		1		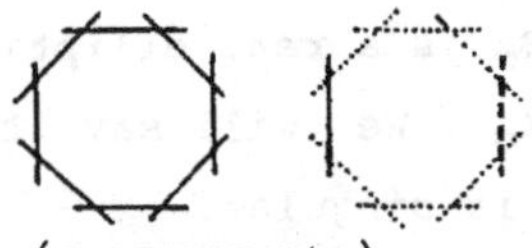 (m components)	-m or 0
m odd	I_m m odd		*	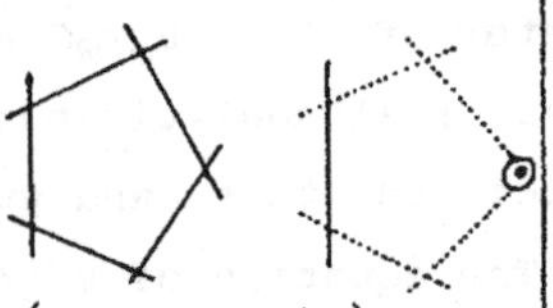 (m components)	-m or 1
$c_1 \begin{cases} v(j) \geq 0 \\ v(\Delta) \equiv 2 \end{cases}$ (mod.12)	II	1			0

$c_2 \begin{cases} v(j) \geq 0 \\ v(\Delta) \equiv 3 \end{cases}$ (mod. 12)	III		*		−1
$c_3 \begin{cases} v(j) \geq 0 \\ v(\Delta) \equiv 4 \end{cases}$ (mod. 12)	IV	1		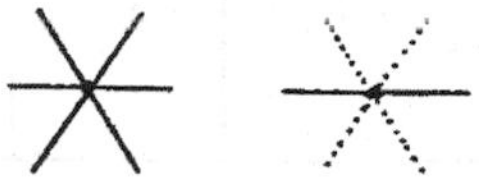	−2 or 0
$c_4 \begin{cases} v(j) \geq 0 \\ v(\Delta) \equiv 6 \end{cases}$ (mod. 12)	I_0^*	1		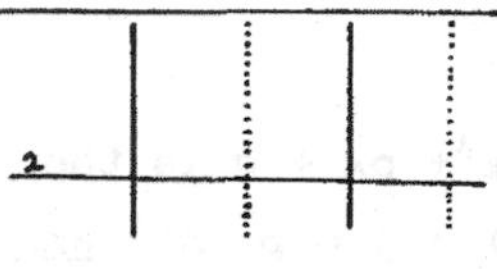	−2
		2		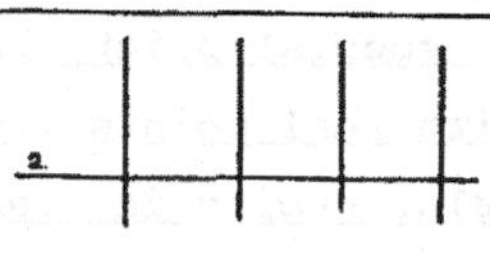	−4
$c_{5_m} \begin{cases} v(j) = -m < 0 \\ m \text{ even} \\ v(q) \text{ odd} \end{cases}$	I_m^* m even	1		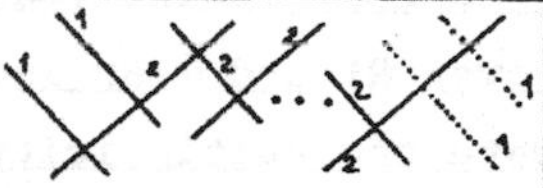 (m+1 double components)	−m−2
		2		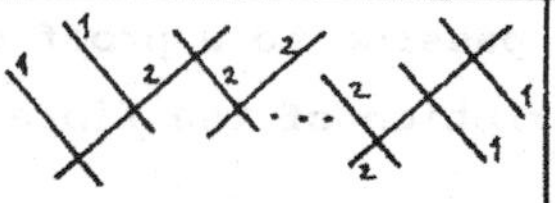 (m+1 double components)	−m−4
$c_{5_m} \begin{cases} v(j) = -m < 0 \\ m \text{ odd} \\ v(q) \text{ odd} \end{cases}$	I_m^* m odd		*	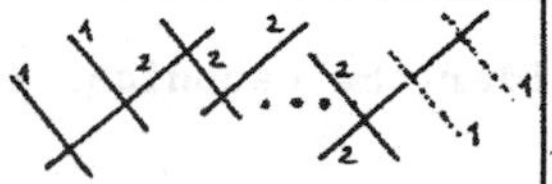 (m+1 double components)	−m−2
			*	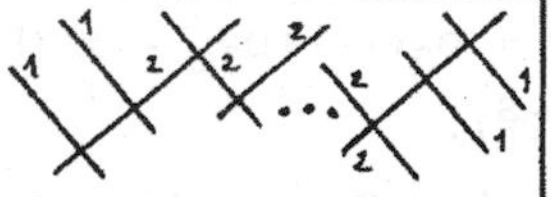 (m+1 double components)	−m−4
$c_6 \begin{cases} v(j) \geq 0 \\ v(\Delta) \equiv 8 \end{cases}$ (mod. 12)	IV^*	1		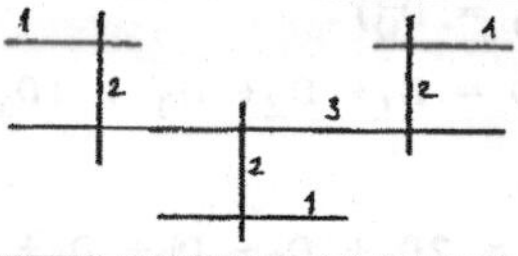	−6
		1		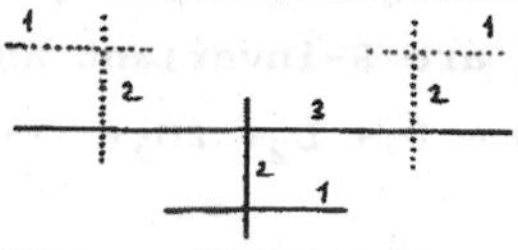	−2

c_7 $\begin{cases} v(j) \geq 0 \\ v(\Delta) \equiv 9 \\ (\text{mod}.12) \end{cases}$	III^*		*	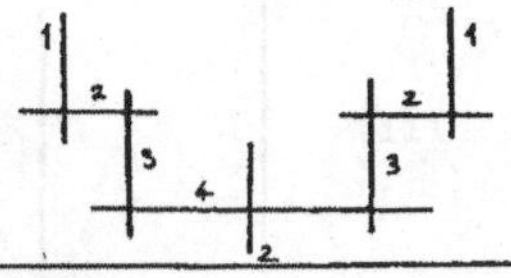	-7
c_8 $\begin{cases} v(j) \geq 0 \\ v(\Delta) \equiv 10 \\ (\text{mod}.12) \end{cases}$	II^*	1		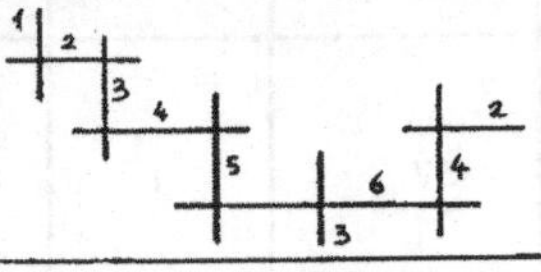	-8

where : $y^2 = x^3 + px + q$ is the Weierstrass equation of the generic fibre over $\mathbb{R}(B_o)$; $j = p^3.\Delta^{-1}$ and $\Delta = 4p^3 + 27q^2$; v is the valuation of $\mathbb{R}(B_o)$ at the special point b_0 ; finally in column 5 —— denotes a real curve with real points (real lines in cases c_2 to c_8 and b_m, $m > 1$, real singular cubics in cases c_1 and b_1) ; ------ a line defined over $\mathbb{R}$ with no real point (case b_m, m even) ; ······ a complex line (not defined over $\mathbb{R}$) and ⊙ an isolted real point (case b_m, m odd). the indices indicate the multiplicity if superior to one.

Before passing to a proof of theorem (1.5) we give here an additional description of the fibre $X_{b_0}(\mathbb{C})$ considered as a divisor on $X = X_o \times_{\mathbb{R}} \mathbb{C}$:

We consider the canonical real structure (X,S) on X and write $D = X_{b_0}(\mathbb{C})$.

Case b_{2m} : $D = D_0 + \cdots + D_{2m-1}$; $(D_i . D_{i+1}) = 1$, $D_i = D_i^S$ for all i or $D_i^S = D_{2m-i}$ $(D_0 = D_0^S)$.
Case b_{2m+1} : $D = D_0 + \cdots\cdot + D_{2m}$; $(D_i . D_{i+1}) = 1$, $D_i = D_i^S$ for all i or $D_i^S = D_{2m+1-i}$ $(D_0 = D_0^S)$.
Case c_3 : $D = D_1 + D_2 + D_3$; $(D_i . D_j) = 1$, $D_i = D_i^S$ for all i or $D_1 = D_1^S$ and $D_2^S = D_3$.
Case c_4 : $D = 2D_0 + D_1 + D_2 + D_3 + D_4$; $(D_0 . D_i) = 1$, $D_i = D_i^S$ for all i or D_0, D_1 and D_3 are S-invariant and $D_2^S = D_4$.
Case c_{5_m} : $D = D_1 + D_2 + 2D_3 + \cdots + 2D_{m+3} + D_{m+4} + D_{m+5}$; $D_i = D_i^S$ for all i

or $D_i = D_i^S$ for $i \leq m+3$ and $D_{m+4}^S = D_{m+5}$.
Case c_6 : $D = D_1 + 2D_2 + 3D_3 + 2D_4 + D_5 + 2D_6 + D_7$; $D_i = D_i^S$ for all i or $D_1^S = D_7$ and $D_2^S = D_6$, the other components being S-invariant.
In cases c_1, c_2, c_7 and c_8 the prime components of the divisor are always S-invariant.

<u>Proof</u> of (1.5) : We start by noting that by V.(1.6) the B_o-minimality of X_0 implies the B-minimality of $X = X_0 \times_{\mathbb{R}} \mathbb{C}$ ($B = B_o \times_{\mathbb{R}} \mathbb{C}$). The section $s : B_o \longrightarrow X_0$ inducing a section $s_{\mathbb{C}} : B \longrightarrow X$ we can use Néron's (or Kodaira's) classification of the special fibres of $\pi_{\mathbb{C}} : X \longrightarrow B$ (see [Né] or [Ko] II). This is essentially the content of <u>columns</u> 1 <u>and</u> 2.

By (1.4) the number of connected components of $\pi^{-1}(b)$, $b \in U$ changes at b_0 if and only if $v(\Delta) \equiv 1 \pmod{2}$. In other words the classification given in column 1 implies the assertions of <u>column</u> 4 in all cases except b_m (I_m) and c_{5_m} (I_m^*). In these last two cases we need to prove that the parity of $v(\Delta)$ only depends on the parity of m. For this we note that in both cases we have by definition of j :

$$-m = v(j) = 3v(p) - v(\Delta)$$

or

$$v(\Delta) = 3v(p) + m . \qquad (1.6)$$

Here we could conclude directly from the fact that cases b_m and c_{5_m} are characterized, respectively, by $v(j.\Delta) \equiv 0 \pmod{12}$ and $v(j.\Delta) \equiv 6 \pmod{12}$ (see [Né] remarque p. 100). But for further use we are going to argue directly using the fact that $m > 0$.

By definition of j, $v(j) < 0$ implies : $3v(p) < v(4p^3+27q^2)$ and hence : $\mathrm{Inf}(3v(p),2v(q)) < v(\Delta)$ which in turn implies : $3v(p) = 2v(q)$. This proves that, in this case, $v(p)$ must be even and, by (1.6), that

$$v(\Delta) \equiv m \pmod{2}. \qquad (1.7)$$

The corresponding results of <u>column</u> 4 follow again from (1.4).

<u>Column</u> 3 : (i) <u>cases</u> c_1, c_3, c_6 <u>and</u> c_8. We have $v(j) = 3v(p) - v(\Delta) \geq 0$. We note that in the cases under consideration, because for any integer m, $3m \neq 2, 4, 8$ or $10 \pmod{12}$, we must in fact have $v(j) > 0$ or $3v(p) > v(\Delta)$ and hence $2v(q) < 3v(p)$.

Let t be a local parameter in the neighbourhood of b_0. Write $q = t^n q'$ and $p = t^m p'$ with $v(q') = v(P') = 0$. We then have :

$$\Delta = t^{2n}(4(t^{3m-2n}p'^3) + 27q'^2)$$

and by the above : $3m-2n > 0$. It follows, that on a sufficiently small neighbourhhod of b_0 , Δ is of the sign of q'^2 and hence positive. The corresponding results on the number of connected components of $\pi^{-1}(b)$ follow from (1.3).

(ii) <u>cases</u> b_m (m <u>even</u>), c_4 <u>and</u> c_{5_m} (m <u>even</u>). It is easy to construct examples corresponding to each of the cases. For b_m (m even) and 1 component we can take $p = (t^m-3)$ and $q = (t^m+2)$. We have $p(0) \neq 0$ so $v(j) = -v(\Delta)$; also, $q(0) \neq 0$ so $v(q) = 0$ is even. For Δ we have $\Delta = 4(t^m-3)^3 + 27(t^m+2)^2 = 8\times 27t^m$ + (terms of higher order). Hence $v(\Delta) = m$ and since m is even, $\Delta \geq 0$ in a neighbourhood of b_0 ($t = 0$). So by (1.3), $\pi^{-1}(b)$ has 1 component in a neighbourhood of b_0. For b_m and 2 components, let $m = 2n$ take $p = t^n-3$, $q = 2-t^n$. This case is very similar to the preceding one except that $\Delta = t^{2n}(4t^n-9)$ and is negative in a neighbourhood of b_0 ($t = 0$).

For the other cases the computations are very much the same, one can take for example, for c_4 and 2 components : $p = -t^2$ and $q = t^6$ or for 1 component : $p = t^2$ and $q = t^3$; for c_{5_m} and 2 components $p = -3t^2$ and $q = t^{m+3} + 2t^3$ or for one component : $p = -3t^2$ and $q = t^{m+3} - 2t^3$.

The results of <u>column</u> 6 will obviously follow from those of column 5. proving the latter will occupy the rest of this section.

<u>Column</u> 5 : To prove the assertion in this column we first note the essential fact, that Néron's theory does not just imply the existence of a regular B_0-minimal model but also the existence on the scheme

$(Reg(\pi^{-1}(b_0))$ of an abelian scheme structure over $\mathbb{R}(b_0) = \mathbb{R}$ (the residue field at b_0). This in turn implies that at least the origin of this group is a smooth real point in the fibre or otherwise said that each fibre as at least one real point not contained in the intersection of two components and not contained in a multiple component. Using the fact that if a real curve has a smooth real point, then it has a Zariski dense subset of real points (see I.(1.14)) we can reformulate this :

(1.8) Under the hypotheses of (1.5), each fibre of π has at least one real component of multiplicity one, with real points.

We prove the results of column 5 case by case. c_1 is trivial. For c_2 we note that there is only one point of intersection between the two components, in particular this point must be real. Since this point is on each of the components a smooth point, I.(1.14) implies that the only possible case is the one described.

For c_7 we note that by (1.7), at least one of the components of multiplicity one must be real. But since there are, by Néron's (or Kodaira's) classification, only two components with multiplicity one, this implies that both must be real. Also, since two distinct components of the fibre have at most one point of intersection we can argue as in case c_2 and prove that again the only possibility is the one described.

The same argument as above also proves the result for c_8.

For c_3, we are going to prove that we can obtain examples for both situations (we note that, trivialy, the two cases described are the only possible). For this we consider first $p = 0$ and $q = t^2$. This gives for the Weierstrass equation $y^2z = x^3+t^2z^3$. The surface we are dealing with is the B_o-minimal regular model of the surface which has this equation in $\mathbb{P}^2\times(D\backslash b_0)$ (D as in (1.5)). We know that such a model exists and that it is unique. This means that one can transform birationally the surface over $D\backslash b_0$ without changing X_0 (or (X,S)).

We consider the automorphism of $\mathbb{P}^2\times(D\backslash b_0)$ defined by $(x,y,z;t) \longmapsto (t^{-1}x,t^{-1}y,z;t)$, t a local real parameter at b_0. Under this transformation the equation defining birationally the surface becomes :

$$y^2z = tx^3 + z^3.$$

For this surface, the fibre over b_0 (corresponding to t = 0) has for equation $z(y^2-z^2) = 0$. Because we are in case c_3 ($v(j) \geq 0$, $v(\Delta) \equiv 3$ mod. 12), this is by Néron [Né] p. 106-107, exactly the equation of the fibre over b_0 of X_0, the smooth B_o-minimal model. This fibre decomposes into 3 real lines defined respectively by : $z = 0$, $y = z$ and $y = -z$. For the other case we can proceed in the same way starting with $p = 0$ and $q = -t^2$. We then obtain for the singular fibre, $z(y^2+z^2) = 0$ or, in other words, the lines defined by $z = 0$, $y = iz$ and $y = -iz$.

<u>Case</u> c_4 : Let again D be as in the theorem and t the associated local parameter. Assume that X_0 is locally defined in $\mathbb{P}^2\times(D\backslash b_0)$ by the equation

$$y^2z = x^3 + p(t)xz^2 + q(t)z^3 .$$

$(x,y,z;t) \longmapsto (t^{2m}x,t^{3m}y,z;t)$, $m \in \mathbb{Z}$, is an automorphism of $\mathbb{P}^2\times(D\backslash b_0)$. The image of the surface under this automorphism is defined by :

$$y^2z = x^3+t^{4m}p(t)xz^2+t^{6m}q(t)z^3 .$$

Reducing to such a model, if necessary, we may always assume that we have in fact $v(\Delta) = 6$. By an argument similar to the one used to establish (1.7) this implies that we must have $v(q) \geq 3$ and $v(p) \geq 2$.

We consider the automorphism of $\mathbb{P}^2\times D'$ (where $D' = D\backslash b_0$) defined by : $(x,y,z;t) \longmapsto (x,y,tz;t)$. The surface obtained by this transformation has for equation :

$$y^2z = t(x^3+t^{-2}p(t)xz^2+t^{-3}q(t)z^3).$$

By the remark made above on $v(p)$ and $v(q)$, this equation defines in fact a surface in $\mathbb{P}^2 \times D$. We consider this surface. The fibre over b_0 (corresponding to $t = 0$) is composed of two lines, one double defined by $y^2 = 0$, the other simple defined by $z = 0$. This surface is singular with singular points : $(\xi_i, 0, 1; 0)$, $i = 1,2$ or 3 where the ξ_i's are the distinct roots of the equation $x^3+t^{-1}p(t)x+t^{-3}q(t) = 0$ (these roots are distinct because the discriminant of the equation is $t^{-6}\Delta$, which, by construction, has valuation 0). The ξ_i's are all three real if $t^{-6}\Delta < 0$ and only one is, if $t^{-6}\Delta > 0$. Since the non-singular minimal model of this surface is, by Néron's classification, obtained by blowing up these three points we find the singular fibres indicated in the table. To show the correspondence with the results of column 3 we use again lemma (1.3) and note, that $t^{-6}\Delta$ is of the same sign as Δ.

Case c_6 : We note that by (1.8), the two forms listed in the table of (1.5) are the only possible. Explicitly one of the lines of multiplicity one must be real, with real points (by (1.8)). Since this line has a unique point of intersection with a line of multiplicity 2, this line also must be real with real points and, for the same reason, also the line of multiplicity 3.

To prove the existence the method is essentially the same as the one used in case c_3, the computations are just that much longer. The skeptic, and courageous, reader can follow the construction given by Néron [Né] p. 112-115, and note that the difference between the two forms comes from the trivial fact that a real polynomial of degree 2 with distinct roots, can have two real roots or two complex conjugate roots.

For the remaining cases, that is b_m and c_{5_m}, the fact that the only possible cases are the one listed is once more a consequence of (1.8), using the fact that if a real line (S-invariant) intersects a complex line it also intersects its complex conjugate or the fact that if two complex conjugate lines intersect in only one point they must intersect in a real point (case b_m, m odd). So we only need to prove the existence and the correspondence with results of column 3.

<u>Case</u> c_{5_m} (m <u>even</u>) : We will need a detailed analysis of the construction given by Néron [Né] p. 108-112.

We recall briefly the main points of this contruction. First one can find a so-called "b_0-standard" equation for the surface over $D\backslash b_0$ (see [Né] proposition 4, p. 95), given by :

$$(1.9)\qquad y^2z + \lambda xyz + \mu yz^2 = x^3 + \alpha x^2z + \beta xz^2 + \gamma z^3$$

with

$$(1.10)\qquad v(\lambda) \geq 1,\ v(\mu) \geq n+2,\ v(\beta) \geq n+2,\ v(\alpha) = 1,$$
$$v(\gamma) \geq 2n+3,\ v(\beta^2-4\alpha\gamma) = 2n+4 \text{ and } 2n = m = -v(j)$$

(see [Né] proposition 6, p. 97).

We proceed as in case c_3, but this time with the isomorphism $(x,y,z;t) \longmapsto (x,y,t^{(n+1)}z;t)$. The new equation is :

$$y^2z + \lambda xyz + t^{-(n+1)}\mu yz^2 = t(t^nx^3 + t^{-1}\alpha x^2z + t^{-(n+2)}\beta xz^2 + t^{-(2n+3)}\gamma z^3)$$

We note that by (1.10) this equation defines a surface in $\mathbb{P}^2\times D$. The fibre over b_0 (corresponding to $t = 0$) is defined by $y^2z = 0$. The surface has 3 singular points : $(1,0,0;0)$, $(\xi_1,0,1;0)$ and $(\xi_2,0,1;0)$, where ξ_1 and ξ_2 are the roots of $t^{-1}\alpha x^2 + t^{-(n+2)}\beta x + t^{-(2n+3)}\gamma_{|t=0}$ which are distinct since they are roots of a second degree equation with discriminant $t^{-(2n+4)}(\beta^2-4\alpha\gamma)_{|t=0}$, which, by (1.10), has valuation zero.

Note also that the points are complex conjugate if $t^{-(2n+4)}(\beta^2-4\alpha\gamma) < 0$ and both real if $t^{-(2n+4)}(\beta^2-4\alpha\gamma) > 0$.

Blowing up once each of the singular points will yield a fibre of the form :

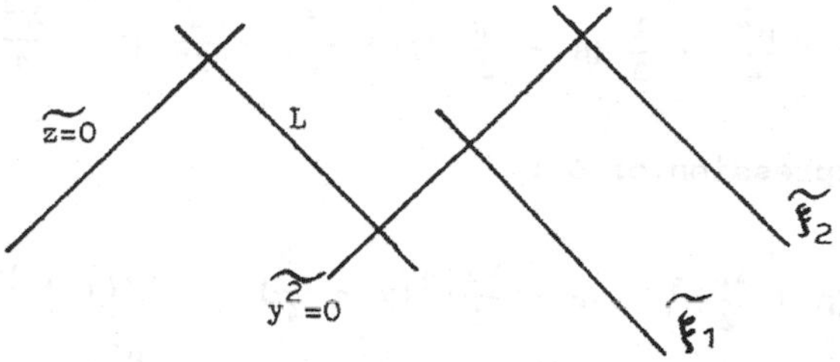

where the components are double components except for the ones denoted $\widetilde{z=0}$, $\widetilde{\xi_1}$ and $\widetilde{\xi_2}$ which correspond to the strict transform of the line z = 0 and to the exceptional lines associated to the points $(\xi_i,0,1;0)$.

Comparing with the final form of the fibre of the Néron model, which is obtained by blowing up additional points of L (some eventually infinitely near)

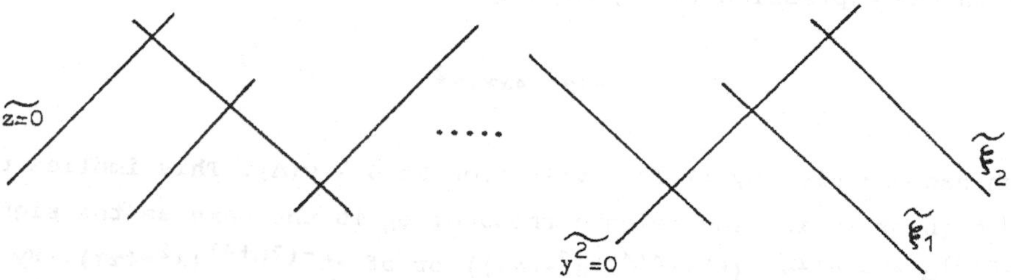

it is easy to see that the two forms of (1.5) only depend on the sign of $t^{-(2n+4)}(\beta^2-4\alpha\gamma)$ at t = 0 and that one can always choose α, β and γ so as to obtain these forms.

We still have to show the correspondence with the results of column 3.

For this we note that to put the equation (1.9) in Weierstrass form, one must take :

(1.11) $\quad p = (\beta + \frac{\lambda\mu}{2}) - \frac{1}{3}(\alpha + \frac{\lambda^2}{4})^2$

$$q = (\gamma + \frac{\mu^2}{4}) - \frac{1}{3}(\beta + \frac{\lambda\mu}{2})(\alpha + \frac{\lambda^2}{4}) + \frac{2}{27}(\alpha + \frac{\lambda^2}{4})^3 .$$

This gives for the expression of Δ :

$$\Delta = 4(\beta + \frac{\lambda\mu}{2})^3 - (\beta + \frac{\lambda\mu}{2})^2(\alpha + \frac{\lambda^2}{4})^2 + 27(\gamma + \frac{\mu^2}{4})^2 - 18(\gamma + \frac{\mu^2}{4})(\beta + \frac{\lambda\mu}{2})(\alpha + \frac{\lambda^2}{4}) + 4(\gamma + \frac{\mu^2}{4})(\alpha + \frac{\lambda^2}{4})^3 \tag{1.12}$$

By (1.10) we find $v(p) = 2v(\alpha) = 2$. On the other hand, since we are in case c_{5_m}, $v(j) = -m$. This means $v(\Delta) = m + 6 = 2n + 6$.

Noting that the conditions (1.10) imply :

$$v(\beta) = n + 2 \text{ or } v(\gamma) = 2n + 3$$

it is an elementary computation to see that the term of smallest valuation in the expression (1.12) of Δ is :

$$-(\beta^2-4\alpha\gamma)\alpha^2$$

which has exactly, by (1.10), valuation $2n+6 = v(\Delta)$. This implies that the sign of Δ in the neighbourhood of b_0 is the same as the sign of $t^{-(2n+6)}\Delta = -(t^{-1}\alpha)^2(t^{-(2n+4)}(\beta^2-4\alpha\gamma))$ or of $-t^{-(2n+4)}(\beta^2-4\alpha\gamma)$. By the above differenciation of cases the result follows once more from lemma (1.3).

<u>Case</u> c_{5_m} (m <u>odd</u>) : Here we only have to prove the existence. The method is quite similar to the one used in case m even.

If we write the equation of the surface in its "b_0-standard" form (1.9), we have :

$$v(\lambda) \geq 1 , v(\mu) \geq n+1, v(\alpha) = 1, v(\beta) \geq n+2 , v(\gamma) \geq 2n+2 \text{ and } v(\mu^2+4\gamma) = 2n+2 \text{ where } 2n-1 = m . \tag{1.13}$$

(see [Né] proposition 6, p. 97).

We use the same isomorphism as before, $(x,y,z;t) \longmapsto$

$(x,y,t^{n+1}z;t)$ but we write the resulting equation in a slightly different form :

$$y^2z + t^{-(n+1)}\mu yz^2 - t^{-(2n+2)}\gamma\, z^3$$
$$= t^{n+1}x^3 + \alpha x^2z + t^{-(n+1)}\beta xz^2 - \lambda xyz.$$

We note, as before, that this equation defines a surface in $\mathbb{P}^2\times$ D. The singular fibre over b_0 is composed of 3 lines defined by $z = 0$, $y = \xi_1 z$ and $y = \xi_2 z$ where ξ_1 and ξ_2 are the distinct roots of $y^2 + t^{-(n+1)}\mu y - t^{-(2n+2)}\gamma$. (distinct because $v(t^{-(2n+2)}(\mu^2-4\gamma)) = 0$). Following Néron's construction, [Né] p. 109, these 3 lines correspond to 3 of the 4 simple components of the final form of the fibre over b_0 of the regular B_0-minimal model. In other words an adequate choice of μ et γ will lead to the two forms of (1.5).

<u>Case</u> b_m (m <u>even</u>) : In this case we can choose a "b_0-standard" form for the equation of X over $D\backslash b_0$ of the form (1.9) with :

(1.14) $v(\mu) > m/2$, $v(\beta) > m/2$, $v(\gamma) = m$, $v(\alpha) \geqslant 0$, $v(\lambda) \geqslant 0$, and $v(4\alpha + \lambda^2) = 0$

(see [Né] proposition 5, p. 96).

We use the isomorphism $(x,y,z;t) \longmapsto (x,y,t^nz;t)$ $(2n = m)$. The equation thus obtained, defines a surface over D which has a singular fibre over b_0, defined by the equation :

$$y^2z + \lambda xyz - \alpha x^2z - t^{-m}\gamma z^3 = 0 \ .$$

This fibre is composed of a line, defined by $z = 0$ and a conic of equation $y^2+ \lambda xy - \alpha x^2 - t^{-m}\gamma z^2 = 0$. This conic is non degenerate since $v(t^{-m}\gamma) = 0$. The final form of the Néron model is obtained by blowing up a certain number of times the points of intersection of the line and the conic (see [Né] p. 103-105). These two points are real, if and only if $4\alpha + \lambda^2 > 0$ at $t = 0$. In this case all the components of the fibre will be real and the fibre will be of the form :

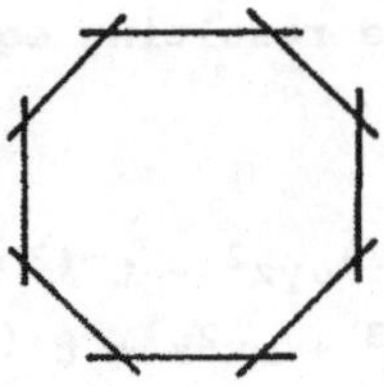

(m components)

If $4\alpha + \lambda^2 < 0$ the points will be complex conjugate and we have two possibilities. In the first, corresponding to $t^{-m}\gamma > 0$ at $t = 0$, the conic is real with real points. In this case the fibre will be of the form :

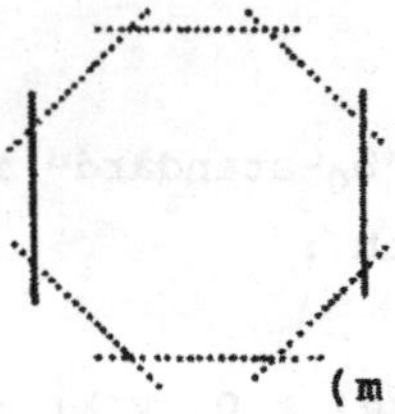

(m components)

or

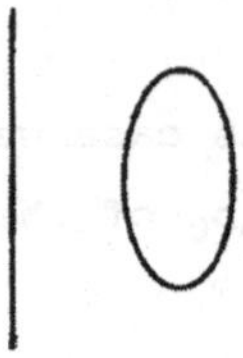

if m = 2 .

If $(4\alpha+\lambda^2) < 0$ and $t^{-m}\gamma < 0$ then the conic is real but has no real points. The form of the fibre of the Néron model will be :

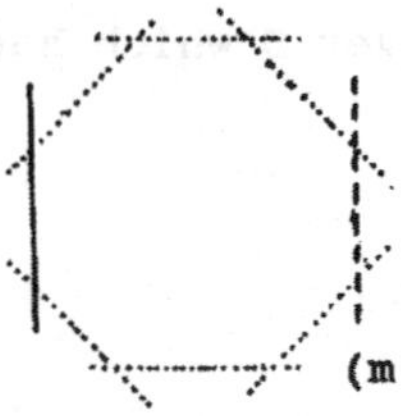

(m components)

or

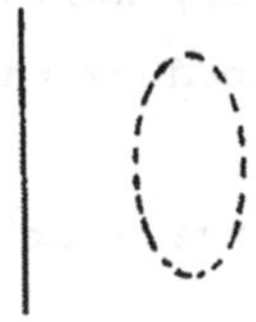

if m = 2 .

To show the correspondence with the results of column 3, we consider once more the expression (1.12) of Δ. The conditions (1.14) imply that the term of smallest valuation in Δ is :

$$4\gamma(\alpha + \frac{\lambda^2}{4})^3$$

and this term is exactly of valuation m. Recalling the sign distinction between the cases, this proves, by lemma (1.3), the assertions.

Case b_m (m odd) : The method of construction is more or less the same as in the case m even. The condition for the "b_0-standard" equation are again given by (1.14). We use the isomorphism $(x,y,z;t) \longmapsto (x,y,\, t^n z;t)$ where $m = 2n+1$. If $m > 1$ (and hence $n > 0$) the equation for the fibre over b_0 is :

$$y^2z + \lambda xyz - \alpha x^2 z = 0$$

In this case the fibre decomposes into 3 lines defined by : $z = 0$, $y = \xi_1 x$ and $y = \xi_2 x$ where ξ_1 and ξ_2 are the distinct roots of $y^2 + \lambda y - \alpha = 0$. These roots are real if $4\alpha + \lambda^2 > 0$ and blowing up the points of intersection of these lines to obtain the Néron model, leads to a fibre of the form :

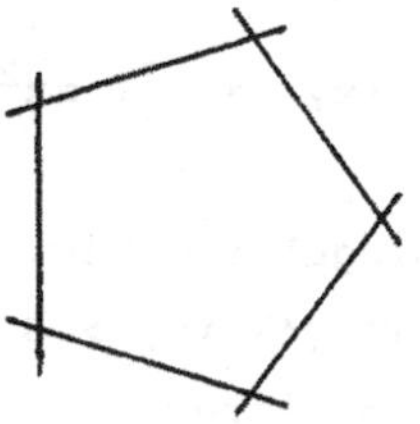

(m components)

If $4\alpha + \lambda^2 < 0$ then the same argument leads to a fibre of the form :

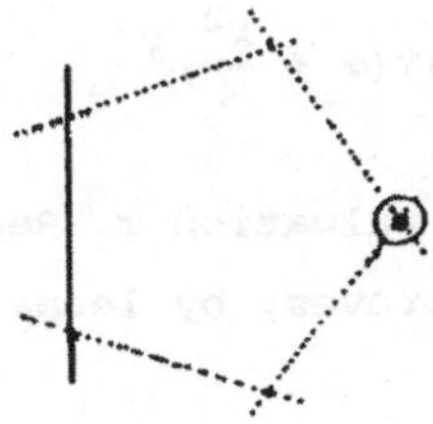

(m components).

To see that there is always an isolated real point in this case, recall the description in terms of divisors. By (1.8) at least one component, say D_0 is fixed under the action of S, with real points. If the action of S does not leave each of the components of the fibre fixed, then elementary considerations concerning the points of intersections show that the only possibility is the one described just after (1.5). In particular we have $S(D_n) = D_{n+1}$ (m = 2n+1) and $D_n.D_{n+1} = 1$. This point of intersection is the isolated real point.

If m = 1 (and n, as above, = 0), then the surface with equation (1.9), satisfying conditions (1.14), is the Néron model. In this case the equation of the fibre over b_0 is :

$$y^2z + \lambda xyz = x^3 + \alpha x^2 z$$

This is the equation of a singular cubic. The singular point (0,0,1) has distinct real tangents if $(4\alpha+\lambda^2) > 0$ and complex conjugate tangents if $(4\alpha+\lambda^2) < 0$, hence the fibres :

or

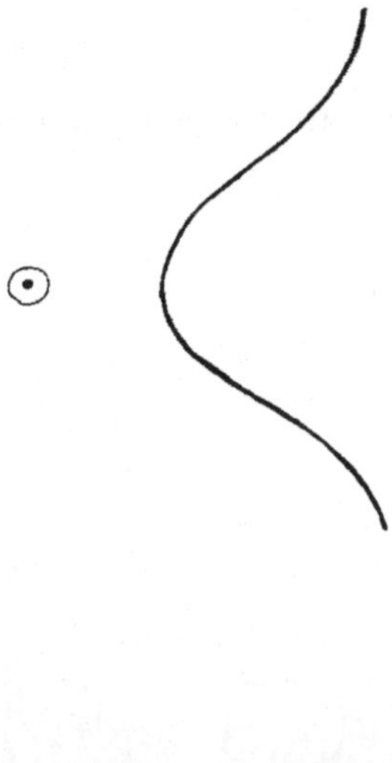

2. Local theory of real elliptic fibrations : the case when there is no section

Let again $\pi : X_0 \longrightarrow B_o$ be a real elliptic fibration. If this fibration does not admit a section, then the first new phenomenon we encounter is that there can exist points $b \in B(\mathbb{R})$ such that $X_b(\mathbb{R}) = \emptyset$. Another phenomenon is that, by Kodaira's classification theorem ([Ko] II, p. 564), we can, in addition to the fibres found in §1, have multiple fibres. These fibres may be of the form μE, where E is a smooth elliptic curve in $X = X_0 \times_{\mathbb{R}} \mathbb{C}$ (we will denote μa such a case) or of the form μL where L is a fibre of the form described in case b_m (we will call μb_m this case).

To specify notations we introduce the following : let Γ be a singular fibre and let $\tilde{\Gamma}$ be the corresponding fibre of the jacobian fibration (see (1.1)) $\tilde{\pi} : \tilde{X}_0 \longrightarrow \tilde{B}_o$ and let $\tilde{\Gamma}$ be of type α ($\alpha = b_m, c_2, \ldots$). If the number of real connected components of the nearby fibres (in the jacobian fibration) is constant and equal to 1 (resp. 2) we will say that Γ is of type $\alpha^{(1)}$ (resp. $\alpha^{(2)}$). If the number of components changes or if there is no ambiguity on the number of components we will say that Γ is of type α.

We return to the notations of §.1. In particular D will be a small disk centred at b_0 and t a local parameter satisfying convention I.(1.11). We will write $\tau = t \circ \pi_{\mathbb{C}}$.

To elliminate one trivial case, we note that if there exists ϵ such that $]-\epsilon, \epsilon[\subset t(D) \cap \mathbb{R}$ and $\tau^{-1}(]-\epsilon,\epsilon[\setminus 0) \cap X(\mathbb{R}) = \emptyset$, then, since X is smooth, $\tau^{-1}(0) \cap X(\mathbb{R}) = \emptyset$.

On the other hand, we note that if for every $t \in \mathbb{R}$ such that $0 < t \leq \epsilon$, $\tau^{-1}(t) \cap X(\mathbb{R}) \neq \emptyset$, then since τ is proper, we have $\tau^{-1}(0) \cap X(\mathbb{R}) \neq \emptyset$.

We are now ready to state and prove the two classifiction theorems for the singular fibres in the case when there is no section.

(2.1) Theorem : Let notations be as above and assume that there exists an ϵ such that $\tau^{-1}(t) \cap X(\mathbb{R}) \neq \emptyset$ for all $t \in \mathbb{R}$ such that $0 \leqslant t < \epsilon$ and $\tau^{-1}(t) \cap X(\mathbb{R}) = \emptyset$ for $-\epsilon < t \leqslant 0$. Then the fibre $\tau^{-1}(0)$ is of type :

- c_2, and the two components of the fibre are complex conjugate. In particular the real part of the fibre is reduced to an isolated point.

- $c_4^{(2)}$ and the fibre is of the form :

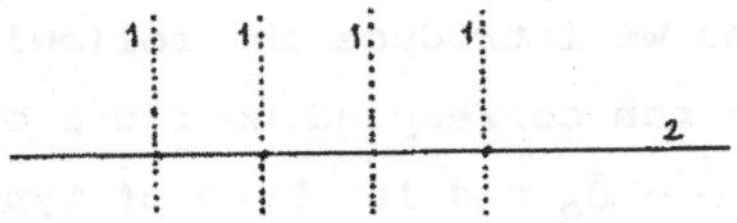

(notations as in theorem (1.5)).

- ${c_5}_m^{(2)}$ and the fibre $\tau^{-1}(0) = \sum_{i=1}^{4} E_i + \sum_{i=0}^{2n} 2D_i$ is of one of the following forms :

$S(D_i) = D_i$ and $D_i(\mathbb{R}) \neq \emptyset$ for all i and $S(E_1) = E_2$, $(E_1.D_0) = (E_2.D_0)$ and $S(E_3) = E_4$;

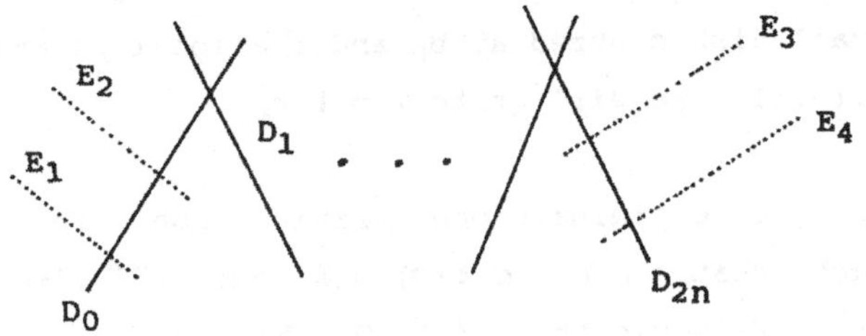

or

$S(D_i) = D_{2n-i}$
$D_n(\mathbb{R}) \neq \emptyset$ and
$S(E_1) = E_3$ or E_4.

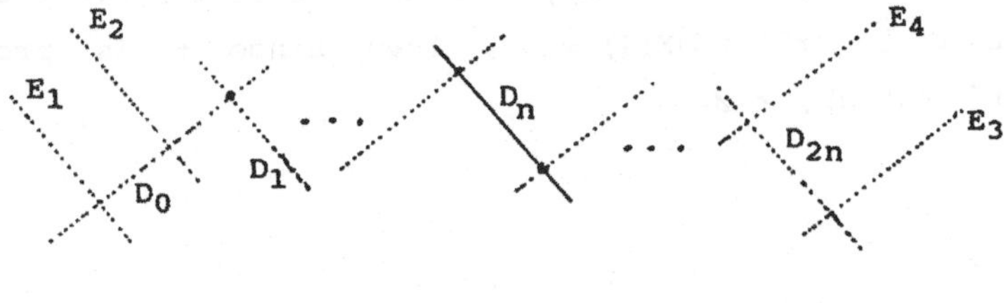

- $c_{5_{2n+1}}$ and the fibre $\tau^{-1}(0) = \sum_{i=1}^{4} E_i + \sum_{i=0}^{2n+1} 2D_i$ verifies one of the following two possibilities :

$S(D_i) = D_{2n+1-i}$

$S(E_1) = E_3,$

$S(E_2) = E_4$;

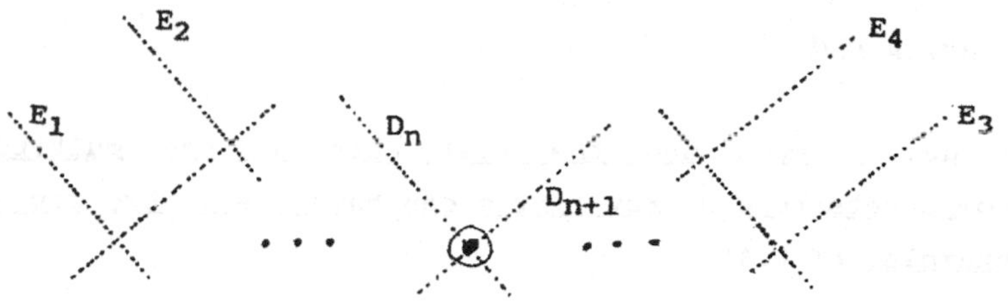

or

$S(D_i) = D_i, \forall\ i,$

$S(E_1) = E_2,$

$S(E_3) = E_4$.

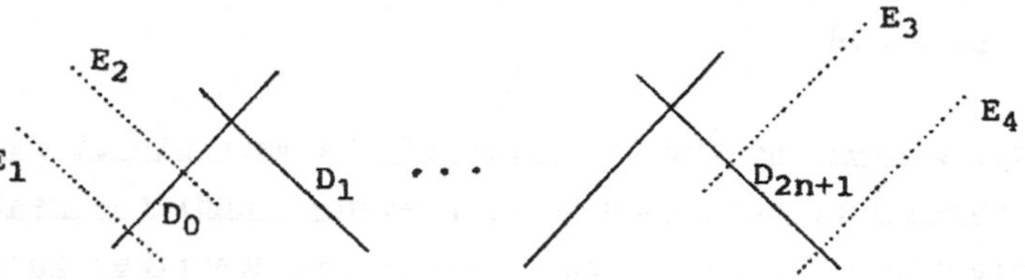

- c_7 and $\tau^{-1}(0)$ is of the form :

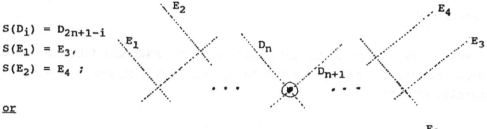

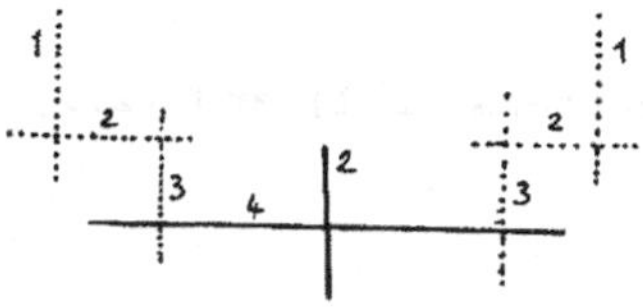

- μa, μ even. For $0 < t < \epsilon$ the real part of the fibres $\tau^{-1}(t)$ always have two components. On the other hand if μE, E irreducible, is the multiple fibre, then $E(\mathbb{R})$ can have one or two components.

- μb_m, μ even and the real part of the fibre is of one of the types described in (1.5).

- μb_m, μ even or odd and the real part of the singular fibre is reduced to one or two isolated points (and this even if $\mu = 1$).

(2.2) Theorem : With the same notations as above, assume that there exists an ϵ such that for all $t \in \mathbb{R}$, $-\epsilon < t < \epsilon$, we have

$\tau^{-1}(t) \cap X(\mathbb{R}) \neq \emptyset$. <u>Then the fibre</u> $\tau^{-1}(0)$ <u>is either of one of the forms described in</u> (1.5) <u>or of one of the following</u> :

- μa, μ <u>odd</u>

- μa, μ <u>even and the real part of the multiple fibre has two components</u> (<u>the nearby fibres can have one or two components</u> - <u>see examples of</u> §.3).

- μb_m, μ <u>odd</u>.

-μb_m, μ <u>even and the real part of the multiple fibre is non-connected</u>. <u>In these last two cases</u> (μb_m, μ <u>even or odd</u>) <u>the real part of the multiple fibre can be of one of the forms described in</u> (1.5) <u>or be composed of exactly</u> 2 <u>isolated points</u> (<u>and this even if</u> $\mu = 1$). <u>In this case we must have</u> $m \equiv 2$ (mod.4).

We are going to prove theorems (2.1) and (2.2) simultaneously.

Let t be a local parameter centred at the special point b_0 and let ϵ be such that $\tau^{-1}(]-\epsilon,\epsilon[)$ (τ as before) contains no singular fibre other than $\tau^{-1}(0)$. We will write $U = \tau^{-1}(]-\epsilon,\epsilon[)$ and $U(\mathbb{R}) = U \cap X(\mathbb{R})$.

We will need :

(2.3) <u>Lemma</u> : <u>Let</u> $\tau^{-1}(0) = Z = \Sigma n_i Y_i$ (<u>where the</u> Y_i's <u>are prime divisors on</u> X) <u>Assume that we are under the hypotheses common to</u> (2.1) <u>and</u> (2.2).

<u>If</u> $U(\mathbb{R})$ (<u>see above</u>) <u>is connected the hypotheses of</u> (2.2) <u>are satisfied if and only if there exists a component</u> Y_i <u>of</u> Z, <u>of odd multiplicity in</u> Z (<u>that is</u> n_i <u>odd</u>) <u>with smooth real points not in any other</u> Y_j.

<u>If</u> $U(\mathbb{R})$ <u>is not connected and the hypotheses of</u> (2.1) <u>are satisfied</u>, <u>then any component</u> Y_i <u>of</u> Z, <u>with a smooth real point not in any</u>

other Y_j, has even multiplicity in Z (that is, n_i is even).

Proof : Assume first that $Z(\mathbb{R})$ is not reduced to a set of isolated points. This implies in particular that there exists a real point x smooth on Y_i and such that $x \notin Y_j$ for $j \neq i$. Note that this implies that Y_i is S-invariant or real, since $S(Y_i)$ is also a component of Z. Assume that there exists such an x on a Y_i, with n_i odd. Let f be a real function defining, in a neighbourhood of x, Y_i in X. Since x is smooth on Y_i, df_x is non zero in $T^*_{X,x}$. Also since x is smooth on X we can find u real in $T^*_{X,x}$ linearly independent from df_x. This fixed, let g be a real function such that $g_x \in m_x \subset O_{X,x}$ and $dg_x = u$. Let C be the curve defined in X by g. By construction C is real and x is a smooth real point of C, in particular $C(\mathbb{R}) \neq \emptyset$. We consider the restriction $\tau_{|C}$ of τ to C. Since C is, by construction, transverse to Y_i and x is not contained in any other Y_j, we have $v_C(\tau_{|C}) = n_i$ (where v_C is the valuation on C centred at x). If n_i is odd, $\tau_{|C}$ (and hence τ) must change sign on $C(\mathbb{R})$, in particular there exists ρ, $\epsilon > \rho > 0$, such that both $\tau^{-1}(\rho)(\mathbb{R})$ and $\tau^{-1}(-\rho)(\mathbb{R})$ are non empty. In other words, the hypotheses of (2.2) are satisfied.

This proves the last assertion of (2.3) and the "if" part of the first assertion.

To prove the "only if" part, let (U_k) be an open covering of $U(\mathbb{R})$ in the ordinary topology. We can always choose (U_k) in such a way that, for all i and k there exists a real function f_{ik} (regular in U_k) defining Y_i in U_k . Let $f_k = \prod f_{ik}^{n_i}$. Since the Y_i's are prime divisors $\tau . f_k^{-1}$ is non zero in U_k. If the n_i's are all even, $f_k \geq 0$ in U_k and if $U(\mathbb{R})$ is connected this implies that τ does not change sign in $U(\mathbb{R})$ (which is smooth). In other words, the hypotheses of (2.1) are satisfied.

To end the proof we consider the case when $Z(\mathbb{R})$ is reduced to a set of isolated points. If $U(\mathbb{R})$ is connected (which is the only case we have to consider), then, $Z(\mathbb{R})$ being the set of real zeros of τ in U, τ cannot change sign in $U(\mathbb{R})$. Hence if $U(\mathbb{R})$ is connected and the points of $Z(\mathbb{R})$ isolated we are again under the hypotheses of (2.1) and this ends the proof.

To prove theorems (2.1) and (2.2) we note that if the special fibre has a non empty real part, then we are under the hypotheses of (2.1) or (2.2) and that these exclude each other.

With this in mind the proof of (2.1) and (2.2) is just a matter of separating cases.

By lemma (1.1) and lemma (1.2) the hypotheses of (2.1), imply that the special fibre $\bar{F}$ of the jacobian fibration must be of one of the following types : $b_m^{(2)}$, b_m with m odd, c_2, $c_4^{(2)}$, $c_{5_m}^{(2)}$, c_{5_m} with m odd or c_7.

By the remark following lemma (1.1), on the jacobian fibration of $X = X_o \times_{\mathbb{R}} \mathbb{C}$, it is enough, to prove (2.1), in the case of non multiple fibres, to study the possible action of S on the above fibres. Applying (2.3) to reduce the possibilities it is elementary, even if painstaking, to prove that the only possibilities are the ones listed in (2.1).

The same argument also proves (2.2) in the case of non multiple fibres.

For fibres of type μa, lemma (2.3), implies that in case (2.1), μ must be even and that if μ is even in case (2.2), then the special fibre has two components.

For fibres of type μb_m, lemma (2.3) implies that in case (2.1) and μ odd, the special fibre can only have isolated real points. A study of the possible actions of S on such a fibre shows that we must have one isolated real point if m is odd, and two isolated real points if m is even. Schematically we can represent the two situations by :

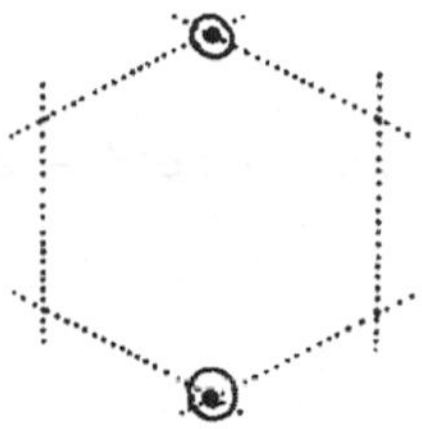 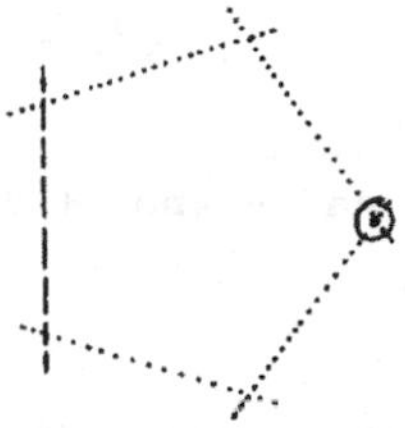

The same argument, shows that in case (2.2), fibre of type μb_m, m even, the real part of the special fibre must be disconnected. A study of the possibilities for the action of S in this case, shows that the singular fibre must have in this case exactly two components. In particular, if the real part has only isolated points, then it has exactly two points.

3. Examples

In this section we give explicit examples for all cases listed in (2.1) and (2.2), except for two cases (see below) the absence of which are most probably due to the authors lack of courage in confrontation to certain computations.

For many constructions we will need :

(3.1) Lemma (Hermite) : *Let C be the affine curve defined over a field K by the equation :*

$$y^2 = ax^4 + 4bx^3 + 6cx^2 + 4b'x + a'.$$

If the polynomial on the right hand side has 4 distinct roots, then the non singular projective completion of C is a curve of genus 1 and its jacobian has for equations :

$$y^2z = 4x^3 - g_2xz^2 - g_3z^3$$

where $g_2 = aa' - 4bb' + 3c^2$ and $g_3 = aca' + 2bcb' - ab'^2 - a'b^2 - c^3$.

For a proof see Weil [We$_3$].

(3.2) example for (2.1) case $c_4^{(2)}$:

We consider the affine surface defined by :

(3.2.1) $$y^2 = t(x^4+1) .$$

We note that if $t < 0$ the fibres over t have no real points, while for $t \geqslant 0$ the real part is non empty. We are hence in case (2.1).

By (3.1) the jacobian fibration associated to this surface is defined by :

$$y^2z = 4x^3 - t^2xz^2 .$$

The j invariant of this jacobian fibration is constant ($\neq 0$) and $\Delta = -t^6$. This proves that we are in case c_4 (and in fact in case $c_4^{(2)}$).

The singular fibre at $t = 0$ defined by the equation (3.2.1) is the double line $y^2 = 0$. This surface has also 4 singular points (all complex) corresponding to the roots of $x^4+1 = 0$. By the same argument as the one used in case c_4 of (1.5), it is easy to show that the fibre of the minimal regular model is of the form indicated in (2.1).

(3.3) examples for (2.1) cases c_2 and c_7 : We apply the method used in case $c_4^{(2)}$ to the affine surfaces defined by :

(3.3.1) $$y^2 = tx^4 - 1$$

and (3.3.2) $$y^2 = tx^4 - t^2$$

We note that in both cases, for $t < 0$, the fibre above t has empty real part.

Applying (3.1) we find in case (3.3.1) : j constant = 0 and $\Delta = t^3$ and for (3.3.2) : j constant = 0 and $\Delta = t^9$. We are hence in cases c_2 for (3.3.1) and c_7 for (3.3.2).

We note that in (3.3.1) the fibre above t = 0 is composed of two complex lines defined by y = i and y = -i and has apparently no real points. But this is because we are dealing with an affine model. The fibre will acquire an isolated real point in the projective completion.

In (3.3.2) we have a double line $y^2 = 0$ and a singular point at y = x = t = 0. Resolving this singular point, will yield another line of multiplicity 4, corresponding to $x^4 = 0$.

(3.4) <u>Examples for</u> (2.1) <u>and case</u> $c_{5_{2n}}^{(2)}$: We consider

(3.4.1) $$y^2 = tx^4 + 6tx^2 + t^{2n+1}$$

and (3.4.2) $$y^2 = t^{2n+1}x^4 + 6tx^2 + t.$$

It is easy to check, using (3.1), that in both examples we are in case (2.1), $c_{5_{2n}}$.

In (3.4.1) the fibre at t = 0 is composed of the double line $y^2 = 0$. The surface has in the fibre, 3 singular points, one real, corresponding to $x^2 = 0$ and two complex, corresponding to $x = \pm i\sqrt{6}$. Comparing with the final form of the non-singular regular model, it is easy to show that this is the case when all the D_i's are real (notations as in (2.1)).

In (3.4.2) the fibre is again composed of the double line $y^2 = 0$ but this time, there are only 2 singular points, defined by $x = \pm i/\sqrt{6}$. This leads to the situation where only D_n (corresponding to $y^2 = 0$) is real.

(3.5) <u>Example for</u> (2.1), <u>case</u> $c_{5_{2n+1}}$: We will only give an example here, for the situation where all the D_i's are real. To obtain this we apply the same method as for (3.4.1) to :

$$y^2 = tx^4 + 6tx^2 - t^{2n+2} .$$

(3.6) <u>Fibres of type</u> μa : Let E_1 be a real elliptic curve, with period matrix $(1,ai)$ ($a \in \mathbb{R}$ and $i^2 = -1$) and let E_2 be the curve with period matrix $(1,½+ai)$. $E_1(\mathbb{R})$ has two components, $E_2(\mathbb{R})$ only one.

E_1 has three non trivial real points of order 2. They are represented in $\mathbb{C}$, by the points : $ai/2$, $1/2 + ai/2$ and $1/2$. E_2 has only one represented by $1/2$.

Let D be the unit disk of $\mathbb{C}$ and let φ be the involution of $D \times E$ ($E = E_1$ or E_2) defined by $\varphi : (z,w) \longmapsto (-z,w+\zeta)$ where ζ is a non trivial point of order 2 on E.

As is well known the quotient X of $D{\times}E$ by φ defines an elliptic fibration over D with a double fibre above 0 (see for example [Gr&Ha] p. 565). Moreover if we choose ζ to be a real point of E, then X has a natural real structure and the fibration is real. We want to study the topology of the real part of X and for this we must differenciate between cases.

We first assume that $E = E_1$ has two real components. In this case we have 3 real points of order 2 to choose from.

a) ζ is the point represented by $ai/2$. The real part of X_0 (the fibre over 0) has 2 components, one obtained by identification of the two components of $E_1(\mathbb{R})$, the other by identification of the points whose representatives in $\mathbb{C}$ are $x+ ai/4$ and $x+ 3ai/4$ ($x \in \mathbb{R}$). If $d \in \mathbb{R}$ and $d > 0$, then the fibre X_d has 2 components, one coming from the identification of points of $D \times E$ of the form (z,x) with $(-z,x+ai/2)$ (z and x real) the other from the identification of $(z,x+ai/2)$ with $(-z,x)$. For $d < 0$, the fibre also has 2 components. One coming from the identification of $(zi,x+ai/4)$ with $(-zi,x+3ai/4)$ (z and x again real), the other from the identification of $(zi,x+3ai/4)$ with $(-zi,x+ai/4)$.

To describe X globally we note that if $d > 0$ tends towards 0, then the two real components of X_d tend towards the same real component of X_0. The same is true if $d < 0$, in other words $X(\mathbb{R})$ has two components (in terms of τ, τ as in (2.1) or (2.2), τ does not change sign on the connected components of $X(\mathbb{R})$, it is positive on one, negative on the other).

b) ζ is the point represented by $(1/2 + ai/2)$. In this case the real part of X_0 has only one component coming from the identification of the two real components of $E_1(\mathbb{R})$ (no other component appears because $x + yi + 1/2 + ai/2 - (x-yi) = 1/2 + (a/2 + 2y)i \notin \mathbb{Z} + a i\mathbb{Z}$). Reasoning as in a) and using the fact noted above, we see that the real part of X_d has 2 components for $d > 0$ and is empty for $d < 0$.

c) ζ is the point represented by 1/2. In this case the real part of X_0 has 2 components, each defined as the image of a component of $E_1(\mathbb{R})$. We find, just as in b), that the real part of X_d has 2 components if $d > 0$ and is empty if $d < 0$.

Topologically $X(\mathbb{R})$ is the disjoint union of two Mobius bands. One is illustrated below :

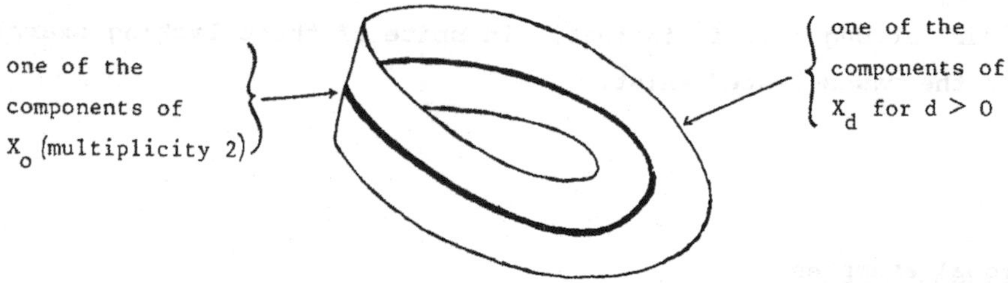

If we start with $E = E_2$, that is, an elliptic curve with only one real component, we only have one non trivial real point of order 2. In this case we have :

d) X_0 has two components, one coming from the real points of E_2, the other from the identification of the points whose representatives are

$x+ ai/2$ and $x+ 1/2 + ai/2$ (to see that these two points are identified, note that in E_2, $S(x + ai/2)$ is represented by $x + 1/2 + ai/2 = x - ai/2 + (1/2 + ai)$). On the other hand, X_d, for $d > 0$ or $d < 0$, has only one component, coming in the first case from the real points of E_2 and in the second from the identification of points of the form $(zi, x + ai/2)$ with $(-zi, x + 1/2 + ai/2)$.

The topological description in this case is the same as in case c).

We end this list of local examples with the example of a fibre of type b_1 and real part reduced to one isolated point (in the next section we will give an example of a fibre of type b_m and real part reduced to two points). For this we consider the surface defined by

$$y^2 = -x^4 + 6(t-1)x^2 + 36t$$

and note that by (3.1) we are in case b_1. This implies that the singular fibre is exactly the one defined by $y^2 = -x^4 - 6x^2$, which has one isolated real point.

This gives examples for all the situations listed in (2.1) and (2.2), except, for one of the cases listed in (2.1), $c_{5_{2n+1}}$ and the cases μb_m.

Our strong belief is that, in spite of these lacking examples, all of the cases listed exist.

4. Global examples

(4.1) An elliptic K3 surface, with real part a torus and two fibres with 2 isolated real points.

We consider the surface defined in $\mathbb{P}^3$ by

$$x^4 - y^4 = z^4 - w^4 .$$

This is the equation of a smooth quartic in $\mathbb{P}^3$, and hence of a K3 surface. It has an obvious elliptic fibration over $\mathbb{P}^1$ whose fibres are the curves defined by

$$\text{(4.1.1)} \qquad \begin{cases} x^2 + y^2 = t(z^2 - w^2) \\ t(x^2 - y^2) = z^2 + w^2 \end{cases} .$$

Since the surface is smooth and minimal, the curves defined by (4.1.1) are the fibres of the regular minimal model. We note that for $|t| < 1$ the fibres have no real points. For $t = 1$ the fibre is formed of 4 lines defined by $x = \pm z$ and $y = \pm iw$. In other words this is a fibre of type b_4. There are 2 real points (1,1,0,0) and (1,-1,0,0).

There are two other real singular fibres, one over -1, one over ∞. For $t = -1$ the situation is the same as the one described above, for $t = 1$. The fibre at infinity is again a fibre of type b_4, but this time all the components are real. With this information at hand it is easy to compute the Euler characteristic, it is equal to 2+2-4 = 0. Since the surface is a quartic in $\mathbb{P}^3$ it is orientable and the real part is hence a torus T_1.

(4.2) A K3 surface with real part connected and of Euler characteristic -18.

We consider in $\mathbb{P}^2 \times \mathbb{P}^1$ ((x,y,z) coordinates of $\mathbb{P}^2$, (t,u) of $\mathbb{P}^1$) the surface defined birationally by

$$\text{(4.2.1)} \qquad y^2 z = x^3 + \frac{(t-u)^5 (t+u)^5}{u^{10}} . z^3 .$$

This defines an elliptic fibration and the j-invariant of this fibration is constant equal to 0. We also have :

$$\Delta = 27 \, \frac{(t-u)^{10} (t+u)^{10}}{u^{20}} .$$

Let X be the regular minimal model of this surface. By the above computation of Δ there are 3 singular fibres, respectively above the points (1,1), (1,-1) and (1,0). For the first two we have $v(\Delta) = 10$. These fibres are hence of type c_8 and by (1.5), the Euler characteristic of the real part of the fibre is in both cases -8. For the fibre at infinity we have, $v(\Delta) = -20 \equiv 4$ (mod.12). The fibre is hence of type c_3. Proceeding as in the proof of (1.5) we can transform the equation (4.2.1) into

$$y^2z = x^3 + u^2(t-u)^5(t+u)^5.z^3$$

or reducing still further, into

$$y^2z = ux^3 + (t-u)^5(t+u)^5z^3 .$$

Just as in the proof of (1.5) we conclude that the fibre of X over (1,0), is defined by $z(y-z)(y+z) = 0$, or in other words, is formed of 3 real lines. The Euler characteristic of the real part of this fibre is -2. This gives for the Euler characteristic of the real part of X, $\chi(X(\mathbb{R})) = -8-8-2 = -18$, as asserted.

It remains to prove that X is a K3 surface. For this we note that by the above computation of the singular fibres, we have

$$\chi(X(\mathbb{C})) = 10 + 10 + 4 = 24 .$$

Since this surface is elliptic, we have for the canonical divisor K, $K^2 = 0$. By Noethers formula this implies that

$$\chi(\mathcal{O}_{X(\mathbb{C})}) = 24/12 = 2 .$$

We also know (cf. [Gr&Ha] ,p. 572) that the canonical divisor K, is of the form $\pi^*(D)$, where D is a divisor on $\mathbb{P}^1$ of degree

$$\deg(D) = 2g(\mathbb{P}^1) - 2 + \chi(\mathcal{O}_{X(\mathbb{C})}) = 0 .$$

Since all divisors of degree 0 on $\mathbb{P}^1$ are trivial we conclude that K = 0. This implies that the Kodaira dimension of the surface is 0, and hence, by the Enriques Kodaira classification of surfaces, that it is a K3 surface.

We also note the following facts on the topology of $X(\mathbb{R})$:

- $X(\mathbb{R})$ is connected. This follows immediately from the fact that the singular fibres are connected.

- Contrary to what was asserted in [Si_3] the surface is orientable. To see this we note that by the preceding description of $X(\mathbb{R})$, $H^1(X(\mathbb{R}),\mathbb{Z}/2)$ is generated by the classes of the irreducible components of the singular fibres, or otherwise said, by the $\pi^*(D)$'s (where the notations are as in III.§1 and $D \in Div(X)$). Since the intersection form on a K3 surface is even, we in particular have, $D^2 \equiv 0$ (mod. 2). By III.(2.5) this implies that for all γ in $H^1(X(\mathbb{R}),\mathbb{Z}/2)$, $\gamma^2 = 0$. By [Mi&Hu] lemma 1.1 p. 101 this in turn implies that $X(\mathbb{R})$ is orientable (see VIII (4.2)).

(4.3) <u>Construction of the example given in</u> III.(4.1).

We consider the elliptic fibration over $\mathbb{P}^1$ defined by

$$y^2z = x^3 + \left(\frac{t^3}{u(t-u)(t+u)}\right)xz^2$$

and take for X the associated minimal regular model. The j-invariant is constant equal to 1 and

$$\Delta = 4\,\frac{t^9}{u^3(t-u)^3(t+u)^3}\ .$$

From this we see that we have 4 singular fibres, above the points, (0,1), (1,0), (1,1) and (-1,1). For each of these fibres we have $v(\Delta) \equiv 9$ (mod. 12). We also have, since j is constant, $v(j) = 0$. By (1.5) this means that the fibres are of type c_7 . If we denote L_i these fibres ($i = 1$ to 4) we have $\chi(L_i(\mathbb{C})) = 9$ and $\chi(L_i(\mathbb{R})) = -7$ and hence $\chi(X(\mathbb{C}) = 36$ and $\chi(X(\mathbb{R})) = -28$. Since by construction $X(\mathbb{R})$ is connected we have $h^1(X(\mathbb{R})) = 30$, as announced.

For the complex part we have, reasoning as in example (4.2),

$$\chi(\mathcal{O}_{X(\mathbb{C})}) = 36/12 = 3.$$

The canonical divisor on X is the pull back of a divisor D on $\mathbb{P}^1$, of degree $d = 2g(\mathbb{P}^1) + \chi(\mathcal{O}_{X(\mathbb{C})}) - 2 = 1$. Since $h^{0,2} = h^{2,0} = \dim H^0(X,\mathcal{O}(K))$ and $K = \pi^*(D)$ we can compute $h^{0,2}$ by Riemann-Roch on $\mathbb{P}^1$. This yields

$$h^{0,2} = d - g(\mathbb{P}^1) + 1 = 2.$$

Since $\chi(\mathcal{O}_X) = 3$, this implies that $h^{0,1} = 0$ and $B_1 = 2h^{0,1} = 0$. Recalling that $\chi(X(\mathbb{C})) = 36$ we then get, $B_2 = 36 - 2 = 34$. This proves all of the assertions relative to example III.(4.1).

(4.4.) <u>Construction of the example given in</u> III.(4.2).

We have the elliptic fibration over $\mathbb{P}^1$, defined birationally in $\mathbb{P}^2 \times \mathbb{P}^1$ by

$$y^2z = x^3 + \left(\frac{u^4}{(u-t)(u+t)(u-2t)(u+2t)}\right) z^3 .$$

There are 5 singular fibres ; 4 for which $\chi(L_i(\mathbb{C})) = 10$ and $\chi(L_i(\mathbb{R})) = -8$ (type c_8) and a 5th for which $\chi(L_5(\mathbb{C})) = 8$ and $\chi(L_5(\mathbb{C})) = -6$ (type c_6). Computations similar to the ones made in (4.3) yield : $B_1 = 0$, $B_2 = 46$, $h^{0,2} = 3$ and $X(\mathbb{R})$ connected, as asserted in III.(4.2).

<u>Bibliographical Notes</u>

Most of this chapter, with the exception of some of the examples is taken from [Si_3], but it is largely inspired by Néron [Né]. Example (4.1) is inspired by an example of Pjateckii-Shapiro and Shafarevich [Pj&Sh], p. 584.

APPENDIX TO CHAPTER VII

REAL HYPERELLIPTIC SURFACES

We give here some indications on how one can study real hyperelliptic surfaces, and treat a few examples.

We first recall the classical description of complex hyperelliptic surfaces (see [B&P&V] p. 147-148, [Be] p. 112-114 or Bombieri, Mumford [Bo&Mu]). Such a surface if of the form $(E \times F)/_{\Gamma}$ where Γ is a sub-group of the group of translation of F and operates on E as a subgroup of the group Aut(E), not contained in the group of translations (for a complete list of possibilites see the references quoted above).

Moreover we have two elliptic fibrations. The first is given by,

$(A_7.1)$ $X = (E{\times}F)/\Gamma \longrightarrow F/\Gamma \cong \mathrm{Alb}(X)$, which is an elliptic curve.

The second by,

$(A_7.2)$ $\qquad X = (E \times F)/\Gamma \longrightarrow E/\Gamma \cong \mathbb{P}^1$

The first has no singular fibres (but is a non trivial fibration). All fibres are in fact, isomorphic to E. The second has for only singular fibres, multiple fibres of type μa (the smooth fibres are isomorphic to F).

If X has a real structure (X,S), then, as noted in V.(1.9)(ii), the fibration $(A_7.1)$ is S-real. In particular, both E and $F/_{\Gamma} = F'$ are real elliptic curves. from this and the above description it is easy to deduce :

(A_7.3) Proposition : A real hyperelliptic surface X, has at most 4 real connected components, depending on the number of components of $E(\mathbb{R})$ and $F'(\mathbb{R})$ (but not only - see below). Moreover each component is homeomorphic to a torns T_1 or a Klein bottle U_2.

To give examples for the above situations, we are going to use the second fibration. For this we assume that both E and F are real elliptic curves and that the action of Γ is compatible with the real structures.

Assume $\Gamma = \mathbb{Z}/2$ in which case Γ operates on F by translation by an element of order 2 and on E by $x \mapsto -x$.

We are going to show that the topology of $X(\mathbb{R})$, $X = (E \times F)/\Gamma$, depends on the number of connected components of $E(\mathbb{R})$ and $F(\mathbb{R})$ and on the choice of the element of order 2.

The map $E \longrightarrow E/\Gamma \cong \mathbb{P}^1$ has 4 ramification points. They correspond to the points of order 2 of E. The 4 are real if $E(\mathbb{R})$ has 2 components, but only 2 are real if $E(\mathbb{R})$ has 1 component (note that the real points of E/Γ come from real and pure imaginary points of E, that is, points such that $S(x) = \pm x$, but for points of order 2 these notions are the same).

Let $\pi : E \times F \longrightarrow (E{\times}F)/\Gamma$ and let e be a real point of order 2 of E. Let t be a real local parameter of E in the neighbourhood of e, satisfying to condition I.(1.11). Then Γ operates on E in the neighbourhood of e by $t \mapsto -t$. In other words, π is in the neighbourhood of e a logarithmic transformation of order 2, exactly as described in VII.(3.6) a), b), c) or d).

From this, we can get an explicit description of the topology of $X(\mathbb{R})$. If the point of order 2 of $F(\mathbb{R})$, which generates Γ, is chosen as in VII.(3.6) a), then locally over a ramification point the surface can be described by

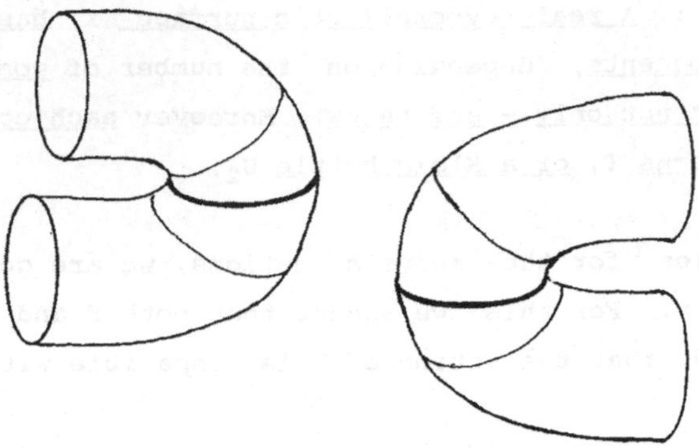

If we have 2 ramification points (resp. 4), $X(\mathbb{R})$ is homeomorphic to $T_1 \amalg T_1$ (resp. $T_1 \amalg T_1 \amalg T_1 \amalg T_1$).

Note that the singular fibre is isomorphic to $2(F/\Gamma) = 2F'$ and that in this case $F'(\mathbb{R})$ has 2 components.

Applying the same procedure to the different choices of the point of order 2 of $F(\mathbb{R})$, we otain $T_1 \amalg T_1$ or T_1 (depending on whether $^{\#}E(\mathbb{R}) = 2$ or 1) if the choice is the one defined in VII.(3.6) b) (note that in this case $^{\#}F'(\mathbb{R}) = 1$) and $U_2 \amalg U_2 \amalg U_2 \amalg U_2$ or $U_2 \amalg U_2$ if the choice is as defined in VII.(3.6) c) or d) (in both these cases $^{\#}F'(\mathbb{R}) = 2$).

We give two more examples. One for which $^{\#}E(\mathbb{R}) = {}^{\#}F'(\mathbb{R}) = 2$ but $^{\#}X(\mathbb{R}) = 2$ or 1.

For this, we choose $\Gamma = \mathbb{Z}/2 \times \mathbb{Z}/2$ and $E(\mathbb{R})$ with period matrix $(1,ai)$. Let the generators of Γ operate on E by $x \mapsto -x$ and $x \mapsto x + \left(\frac{1}{2} + \frac{a}{2}i\right)$. In this case it is easily seen that only 2 of the ramification points of $E \longrightarrow E/\Gamma$ are real (one from the identification of 0 and $\frac{1}{2} + \frac{a}{2}i$ the other from the identification of $\frac{1}{2}$ and $\frac{a}{2}$ i).

Now choose F, such that $^{\#}F(\mathbb{R}) = 2$. In this case, we have $^{\#}F'(\mathbb{R}) = 2$. Let for example F be defined by the period matrix $(1,bi)$. If the generator of Γ, which operates on E by $x \mapsto -x$ is $\frac{1}{2} + \frac{b}{2}$ i, then $X(\mathbb{R})$

has only 1 component ($X(\mathbb{R})$ will have 2 if we choose $\frac{1}{2}$ or $\frac{b}{2}$ i, 2 Klein bottles in the first case, 2 tori in the second).

We give a last example where $X(\mathbb{R})$ has 3 components. For this, we let $\Gamma = \mathbb{Z}/4$ and E defined by the period matrix (1,i). 3 of the ramification points of the map $E \longrightarrow E/\Gamma$ are real (coming from the points 0, $\frac{1}{2} + \frac{i}{2}$ and the identification of $\frac{1}{2}$ and $\frac{i}{2}$). Let F be again defined by the matrix (1,bi), then depending on if we choose the generator of Γ to be $\frac{1}{4}$ or $\frac{b}{4}$ i we obtain 3 tori or 3 Klein bottles.

VIII. REAL K3 SURFACES.

1. Conditions for a K3 surface to be real and real Torelli theorem

For definitions and general results on K3 surfaces our main reference will be the Palaiseau seminar on K3 surfaces [X].

By definition a K3 surface is a complete complex surface with trivial canonical class and first Betti number, $B_1 = 0$. In our treatment of K3 surfaces, we will use of one important fact not contained in the definition, namely, Siu's result that all K3 surfaces are Kähler (see [X] exposé XII).

To homogenize our notations we will write X for a K3 surface, whether it is algebraic or not (refering in the second case to the set of complex points with the complex structure).

We will say that X is real, if it is algebraic and real in the sense of I.(1.15) or if X is only Kähler, if there exists an antiholomorphic involution S on X (since for surfaces complete and algebraic imply projective, these definitions are coherent, by I.(1.3)). In both cases we will speak of the real structure (X,S) (as we have done in II.§1).

We make one last remark on notations, $X(\mathbb{C})$ and X_{top} will both denote the set of complex points, but with special emphasis in the second case on the fact that we only consider the topological structure.

Let X be a complex K3 surface, that we assume for the moment algebraic, and let X^{σ} be the conjugate surface (in the sense of I.§1). By definition, we can consider X^{σ} as the surface obtained by considering on X_{top}, the complex structure conjugate to the one of X. With this in mind we may identify $H^*(X(\mathbb{C}),\mathbb{Z}) = H^*(X_{top},\mathbb{Z}) = H^*(X^{\sigma}(\mathbb{C}),\mathbb{Z})$.

Let $H^*(X(\mathbb{C}),\mathbb{C}) = \oplus H^{p,q}(X)$ (resp. $H^*(X^\sigma(\mathbb{C}),\mathbb{C}) = \oplus H^{p,q}(X^\sigma)$) be the Hodge decomposition of $H^*(X(\mathbb{C}),\mathbb{C})$ (resp. $H^*(X^\sigma(\mathbb{C}),\mathbb{C})$).

With the above identification we can consider both, $H^{p,q}(X)$ and $H^{p,q}(X^\sigma)$ as subspaces of $H^*(X_{top},\mathbb{C})$. In this case, we have, by definition of X^σ :

$$(1.1) \qquad H^{p,q}(X) = \overline{H^{p,q}(X^\sigma)} = H^{q,p}(X^\sigma) \ .$$

Let D be an effective divisor (resp. an ample divisor) on X and let d be its Chern class. Let D^σ be the corresponding divisor on X^σ (see I.(4.13)). Obviously D^σ is also effective (resp. ample) on X^σ. But as noted in I.(4.14) the Chern class in $H^2(X_{top},\mathbb{Z})$ is -d. In other words :

(1.2) _If_ $d \in H^2(X_{top},\mathbb{C})$ _is the Chern class of an effective divisor_ (_resp. an ample divisor_) _on_ X, _then_ -d _is the Chern class of an effective_ (_resp. ample_) _divisor on_ X^σ.

Let A_X (resp. A_{X^σ}) be the sub-set of $H^2(X_{top},\mathbb{Z})$ formed by the classes of effective divisors on X (resp. X^σ). By (1.2) we have :

$$(1.3) \qquad A_X = -A_{X^\sigma} \ .$$

Let $V(X) = \{x \in H^{1,1}(X) \cap H^2(X_{top},\mathbb{R}) \ / \ x^2 > 0\}$ and let $V(X^\sigma)$ be the corresponding cone for X^σ. By (1.1), $H^{1,1}(X) = H^{1,1}(X^\sigma)$, hence

$$V(X) = V(X^\sigma) \ .$$

If $V^+(X)$ is the connected component of $V(X)$ containing the classes of ample divisors (see [X] exposé IV, 64 or exposé VII, §2), then by (1.2) :

$$(1.4) \qquad V^+(X^\sigma) = -V^+(X)$$

where $V^+(X^\sigma)$ is defined in the same way as $V^+(X)$.

(1.5) If X is only a Kähler K3 surface we define A_X and $V(X)$ in the same way but we define $V^+(X)$ to be the component of $V(X)$ containing Kähler classes (see II.§1 or [X] IV). Recalling II.(1.4) we note that (1.4) still holds in this case.

Let A be a $\mathbb{Z}$-module and $\tau : A \longrightarrow A$ a morphism. For K any field, we will write τ_K for $\tau \otimes \mathrm{Id} : A \otimes K \longrightarrow A \otimes K$.

(1.6) Theorem : A K3 surface X is real if and only if there exists an involution τ on $H^2(X(\mathbb{C}),\mathbb{Z})$ such that :

(i) τ is an isometry for the cup-product form ;

(ii) $\tau_{\mathbb{C}}(H^{p,q}(X)) = H^{q,p}(X)$ for $p+q = 2$;

(iii) $\tau(A_X) = -A_X$;

(iv) $\tau(V^+(X)) = -V^+(X)$.

Under these hypotheses there exists a unique real structure (X,S) such that $S^* = \tau$.

Proof : We first check that the conditions are necessary. Let (X,S) be a real structure on S. S^* is an involution on $H^2(X(\mathbb{C}),\mathbb{Z})$ and by II.(1.1), it is an isometry. By definition we have $S^*_{\mathbb{C}} = F_\infty$ and, by I.(2.4), (ii) is verified. Finally (iii) and (iv) follow from I.(4.12).

To prove that the conditions are necessary we are going to use the Torelli theorem for complex K3 surfaces .

By (1.1), (1.3) and (1.4), and the definition of τ, we have, $\tau_{\mathbb{C}}(H^{p,q}(X)) = H^{p,q}(X^\sigma)$, $\tau(A_X) = A_{X^\sigma}$ and $\tau_{\mathbb{R}}(V^+(X)) = V^+(X^\sigma)$. In other words, τ is an effective Hodge isometry between $H^2(X(\mathbb{C}),\mathbb{Z})$ and $H^2(X^\sigma(\mathbb{C}),\mathbb{Z})$ (see for example [X] exposé VII §3). But by the Torelli theorem this means that there exists a unique isomorphism $\varphi : X^\sigma \longrightarrow X$ such that $\varphi^* = \tau$. By the identification as point sets of $X(\mathbb{C})$ and $X^\sigma(\mathbb{C})$ we can consider φ as a map from X to X. In this case φ is anti-holomorphic. Also by the unicity of φ, φ is an involution. This means, by I.(1.3) that φ defines a real structure (X,φ) on X. By construction this is the only one for which $\varphi^* = \tau$.

(1.7) <u>Corollary</u> (Real Torelli theorem for K3 surfaces) : <u>Let</u> X <u>and</u> Y <u>be two</u> K3 <u>surfaces and</u> $\varphi : X \longrightarrow Y$ <u>a complex isomorphism</u>. <u>If</u> X <u>and</u> Y <u>are both real, with real structures</u> (X,S) <u>and</u> (Y,T), <u>then</u> φ <u>is a real isomorphism</u> (<u>for</u> S <u>and</u> T) <u>if and only if</u> φ^* <u>commutes with the Galois actions on</u> $H^2(Y(\mathbb{C}),\mathbb{Z})$ <u>and</u> $H^2(X(\mathbb{C}),\mathbb{Z})$, <u>that is</u> $\varphi^* \circ T^* = S^* \circ \varphi^*$.

<u>Proof</u> : Just apply (1.6) to $(\varphi^{-1})^* \circ S^* \circ \varphi^*$.

We can reformulate this by saying :

(1.8) <u>Corollary</u> : <u>Two real</u> K3 <u>surfaces</u> (X,S) <u>and</u> (Y,T) <u>are real isomorphic if and only if there exists an effective Hodge isometry</u> $f : H^2(X(\mathbb{C}),\mathbb{Z}) \longrightarrow H^2(Y(\mathbb{C}),\mathbb{Z})$ <u>that commutes with the Galois actions</u>.

For later use we introduce some notations and another formulation of (1.6).

Let

$$\Delta(X) = \{x \in H^2(X(\mathbb{C}),\mathbb{Z}) \cap H^{1,1}(X) \ / \ x^2 = -2\}$$

be the set of divisor classes of square -2. We note here, that if e is any divisor class of square -2, then, by Riemann Roch, either e or -e is effective. In particular if $\Delta^+(X)$ is the sub-set of $\Delta(X)$ formed by effective divisor classes, then

$$\Delta(X) = \Delta^+(X) \cup -\Delta^+(X) \quad .$$

Let H_e be the hyperplan in $L_{\mathbb{R}}$, defined by $\{x \in L_{\mathbb{R}} \ / \ \langle x.e\rangle = 0\}$ and let

$$\mathcal{H} = \bigcup_{e\in\Delta(X)} H_e = \bigcup_{e\in\Delta^+(X)} H_e \quad .$$

The hyperplanes H_e cut out chambers in C(X), i.e., connected components in $C(X)\setminus\mathcal{H}$ (see [Bou] chap. V for the terminology). One and only one, of these chambers contains Kähler classes (or in the algebraic case classes of ample divisors). We denote this chamber by $\mathcal{K}_X$. It is characterised by the fact that for any $x \in \mathcal{K}_X$ and any $e \in \Delta^+(X)$, $<x.e> > 0$

(1.9) Conditions (iii) and (iv) of (1.6) can be replaced by $\tau_{\mathbb{R}}(\mathcal{K}_X) = -\mathcal{K}_X$ (see [X] VII §3).

We also introduce the following. For $e \in \Delta(X)$, let $s_e(x) = x + <x.e>e$ be the reflection relative to H_e and let W_X be the group generated by such reflections (note that since $s_e = s_{-e}$, W_X is in fact generated by the s_e's for $e \in \Delta^+(X)$). The main result we will use is :

(1.10) W_X (resp. $W_X \times \{\pm 1\}$) operates simply transitively on the chambers of $C^+(X)$ (resp. $C(X)$) (see [X] VII, §3 or [Do]).

2. Moduli space of real K3 surfaces

By a <u>lattice</u> we shall mean a finitely generated $\mathbb{Z}$-module endowed with an integral, symmetric, non degenerate, bilinear form $< , >$.

If X is a K3 surface, then it is well known (see [X] IV) that $H^2(X(\mathbb{C}),\mathbb{Z})$ with its cup-product form is isomorphic to <u>the lattice</u> L <u>of signature</u> (3,19). L is unique up to isomorphism. In fact L is $H \perp H \perp H \perp (-E_8) \perp (-E_8)$, where H is the hyperbolic lattice of rank 2 (form defined by $\begin{pmatrix} 0 & 1 \\ 1 & 0 \end{pmatrix}$), and $-E_8$ is the negative definite rank 8 lattice associated to the Dynkin diagram :

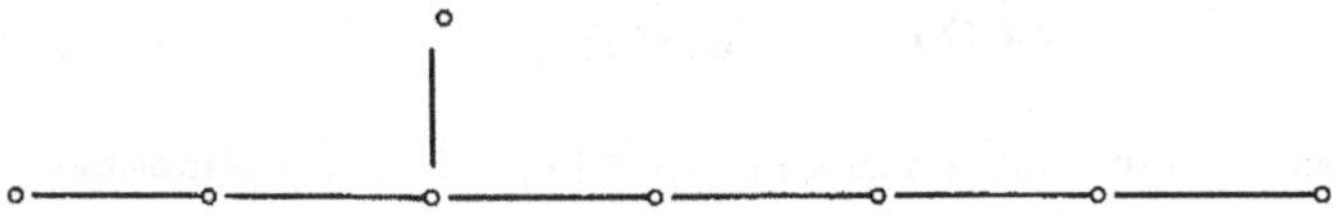

We will say that an isomorphism $f : H^2(X(\mathbb{C}),\mathbb{Z}) \longrightarrow L$ of lattices, is a <u>marking</u> on the K3 surface X and say that (X,f) is a <u>marked</u> K3 surface.

Let (X,f) be a marked K3 surface and let $P \subset L \otimes \mathbb{R}$ be the image under $f_{\mathbb{R}}$ of $H^2(X(\mathbb{C}),\mathbb{R}) \cap (H^{0,2}(X) \oplus H^{2,0}(X))$. Since $H^{2,0}(X)$ is of complex dimension 1, P is a real 2-plane in $L \otimes \mathbb{R}$. We choose an orientation on P in such a way that for any $\omega \in H^{2,0}(X)$, the basis $(\mathrm{Re}(\omega),\mathrm{Im}(\omega))$ is direct.

Recall for later use, that the restriction of the cup-product form to $H^{2,0}(X) + H^{0,2}(X)$ is positive definite and that for any $\omega \in H^{2,0}(X)$, $\mathrm{Re}(\omega)$ and $\mathrm{Im}(\omega)$ are orthogonal (see [X] exposé IV).

<u>We will say that this oriented plane</u> P <u>is the period of</u> (X,f).

Recall that by the complex Torelli theorem, the period P characterizes (X,f) up to isomorphism, or in other words that the period map is injective (see [X] exposé VII).

We will say that an involution τ on $H^2(X(\mathbb{C}),\mathbb{Z})$ induces a real structure on X, if τ satisfies the conditions of theorem (1.6).

(2.1) <u>Lemma</u> : <u>Let</u> (X,f) <u>be a marked</u> K3 <u>surface and let</u> P <u>be its period</u>. <u>Let</u> k <u>be a Kähler class of</u> X <u>and let</u> E <u>be the positive definite</u> 3-<u>plane</u> $P \oplus f_{\mathbb{R}}(k)\mathbb{R}$ <u>in</u> $L \otimes \mathbb{R}$.

<u>Let</u> τ <u>be an involution of</u> L <u>and assume</u> :

(i) E <u>is stable under</u> $\tau_{\mathbb{R}}$;

(ii) <u>The restriction of</u> $\tau_{\mathbb{R}}$ <u>to</u> E <u>is a symmetry relative to a line in</u> P.

<u>Then</u> $f^{-1}\circ\tau\circ f$ <u>induces a real structure on</u> X.

<u>Proof</u> : Write $\tilde{\tau} = f^{-1}\circ\tau\circ f$. The hypothesis that the restriction of $\tau_{\mathbb{C}}$ to E is a symmetry relative to a line of P implies that P is stable under $\tau_{\mathbb{R}}$ and that the restriction of $\tau_{\mathbb{R}}$ is also a symmetry relative to a line. Replacing ω by $\alpha\omega + i\beta\omega$ if necessary we may always assume that the line is generated by $\mathrm{Re}(\omega)$. In this case $\tau_{\mathbb{R}}(\mathrm{Re}(\omega)) = \mathrm{Re}(\omega)$ and $\tau_{\mathbb{R}}(\mathrm{Im}(\omega)) = -\mathrm{Im}(\omega)$. This implies that $\tilde{\tau}_{\mathbb{C}}(H^{0,2}) = H^{2,0}$. Since

$H^{1,1}(X) = (H^{0,2} \oplus H^{2,0})^{\perp}$, we also have $\tilde{\tau}_{\mathbb{C}}(H^{1,1}(X)) = H^{1,1}(X)$ because $\tilde{\tau}$ is by definition an isometry.

To end the proof of the lemma we only need to prove that $\tilde{\tau}$ satisfies conditions (iii) and (iv) of (1.6).

By hypothesis we have,

(2.2) $$\tilde{\tau}(k) = -k \quad .$$

We note that by definition of $V(X)$, $\tilde{\tau}(V(X)) = V(X)$.

Since $V(X) = V^+(X) \cup -V^+(X)$ and x and y belong to the same component, $V^+(X)$ or $-V^+(X)$, if and only if $<x.y> > 0$, then we must have by (2.2), $\tilde{\tau}_{\mathbb{R}}(V^+(X)) = -V^+(X)$.

For A_X, we first note that by Lefschetz's theorem on (1,1)-classes, the set of divisor classes is $H^2(X(\mathbb{C}),\mathbb{Z}) \cap H^{1,1}(X)$. By the above, this means that $\tilde{\tau}$ sends divisor classes to divisor classes.

Let c be the class of an irreducible curve. By adjunction, $c^2 \geqslant -2$ and hence, $(\tilde{\tau}(c))^2 \geqslant -2$. By Riemann-Roch this implies that $\tilde{\tau}(c)$ or $-\tilde{\tau}(c)$ is effective. But for an effective divisor class e, we must have $<e.k> > 0$. By (2.2) $-\tilde{\tau}(c)$ is effective. This proves that $\tilde{\tau}(A_X) = -A_X$.

(2.3) Theorem : Let S be an involution on L and let $L^G = \{\ell \in L \,/ S(\ell) = \ell\}$. There exists a marked K3 surface (X,φ) such that $\tilde{S} = \varphi^{-1} \circ S \circ \varphi$ induces a real structure on X if and only if the restriction of the bilinear form to L^G has signature $(1,b-1)$, where $b = \text{rank } L^G$ (note that by assumption, $b \geqslant 1$).

Proof : We first note that by II.(1.2), the condition is necessary.

Let S be an involution on L satisfying to the conditions of (2.3).

We are first going to show that we can choose an x in $L^G_{\mathbb{R}} = L^G \otimes \mathbb{R} = (L \otimes \mathbb{R})^G$ such that $x^2 > 0$ and such that for any $e \in L^G$, $x \notin e^{\perp}$.

To prove this, we note that if $b = \text{rank } L^G = 1$, then the condition on the signature implies that the restriction of the bilinear form to $L_{\mathbb{R}}^G$ is positive definite. The result is obvious in this case. If $b > 1$, then for any $e \in L^G$, $e \neq 0$, $L_{\mathbb{R}}^G$ cannot be contained in $e^{\perp}$ nor can $L_{\mathbb{R}}^G$ be contained in $\bigcup_i e_i^{\perp}$ for any discrete family $\{e_i\}$. The same is also true for any non empty open set of $L_{\mathbb{R}}^G$. The assertion follows from the fact that, by the assumption on the signature, the set $\{x \in L_{\mathbb{R}}^G / x^2 > 0\}$ is a non empty open cone in $L_{\mathbb{R}}^G$ and that L^G is discrete.

We choose such an x.

Let $(L_{\mathbb{R}}(1))^G = \{\ell \in L_{\mathbb{R}} / S(\ell) = -\ell\}$. We have $(L_{\mathbb{R}}^G) = (L_{\mathbb{R}}(1))^G$. The condition on the signature implies that the restriction of the bilinear form to $(L_{\mathbb{R}}(1))^G$ has signature $(2, r-2)$, where $r = 22-b$.

Let $y \in L_{\mathbb{R}}(1))^G$, such tht $y^2 > 0$ and let P be the plane generated in $L_{\mathbb{R}}$, by x and y and consider on P the orientation defined by (x,y).

We note that the condition on x implies that for any $e \in L \cap P^{\perp}$, $e^{\perp}$ does not contain $(L_{\mathbb{R}}(1))^G$ (if not e would be in $L \cap (L_{\mathbb{R}}(1))^G)^{\perp} = L \cap L_{\mathbb{R}}^G = L^G$ and x in $e^{\perp}$, contrary to hypothesis). Using of the same argument as for x, we can find $k \in y^{\perp} \cap L_{\mathbb{R}}(1)$ such tht $k^2 > 0$ and for any $e \in L \cap P^{\perp}$, with $e^2 = -2$, $k \notin e^{\perp}$.

Under these conditions there exists, by the surjectivity of the period map (see [X] exposé X, p. 127), a marked K3 surface (X,φ) of period P and such that $\varphi^{-1}(k)$ is a Kähler class on X. Since, by construction, the restriction of S to $P \oplus k\,\mathbb{R}$ is a symmetry relative to the line generated by x, the theorem follows from lemma (2.1).

(2.4) <u>Remark</u> : In practice the condition we have imposed on x, in the proof of (2.3), is somewhat too strong (i.e. we would not obtain all real K3 surface in this way). All we realy need is that for any $e \in L \cap P^{\perp}$, such that $e^2 = -2$, we have $e^{\perp} \not\supset (L_{\mathbb{R}}(1))^G$. Since $(L_{\mathbb{R}}^G)^{\perp} = L_{\mathbb{R}}(1)^G$, we can replace this condition by $e \notin L_{\mathbb{R}}^G$. In other words it is enough to choose $x \in L_{\mathbb{R}}^G$, such that $x^2 > 0$ and $x^{\perp} \cap \{e \in L^G / e^2 = -2\} = \emptyset$.

We are now going to describe the space of moduli of real K3 surfaces, or rather the spaces of moduli.

For this we fix an involution S on L, satisfying to the conditions of (2.3).

Let C_1 be the cone in $L_{\mathbb{R}}^G$ defined by $x^2 > 0$. Since by hypothesis, the restriction of the bilinear form to $L_{\mathbb{R}}^G$ has signature $(1,b-1)$, C_1 decomposes into 2 connected components C_1^+ and $-C_1^+$. Let $\tilde{\Omega}_1^+(S)$ be the image of C_1^+ in $\mathbb{P}(L_{\mathbb{R}}^G) = \mathbb{P}^{b-1}(\mathbb{R})$.

Let C_2 be the cone in $L_{\mathbb{R}}(1)^G$, defined by $y^2 > 0$. Since the signature is this time $(2,r-2)$, C_2 is connected. Let $\Omega_2(S)$ be the image of C_2 in the set of oriented lines in $L_{\mathbb{R}}(1)^S$. One can consider $\Omega_2(S)$ as an open set in the hypersphere S^{22-b-1}.

Let $e \in L^G$, such that $e^2 = -2$. The hyperplanes $H_e = \{x \,/\, \langle x.e\rangle = 0\}$ of $\mathbb{P}(L_{\mathbb{R}}^G)$ cut out open sets in $\Omega_1^+(S)$. Let $\tilde{\Omega}_1^+(S)$ be the union of these open sets.

(2.5) <u>Theorem</u> : <u>Let</u> S <u>be an involution on</u> L <u>satisfying the conditions of</u> (2.3). <u>The set of isomorphy classes of marked</u> K3 <u>surfaces</u> (X,φ) <u>on which</u> S <u>induces a real structure is parameterised bijectively by the open set</u> $\Omega(S) = \tilde{\Omega}_1^+(S) \times \Omega_2(S)$ <u>of</u> $\mathbb{P}^{b-1}(\mathbb{R}) \times S^{22-b-1}$.

<u>Proof</u> : Let $x \in \tilde{\Omega}_1^+(S)$ and $y \in \Omega_2(S)$. Let P be the corresponding 2-plane in $L_{\mathbb{R}}$, oriented by $(\tilde{x},\tilde{y})$, where $\tilde{x} \in C_1^+$ is in the inverse image of x and $\tilde{y} \in C_2$ is in the inverse image of y.

By definition of $\tilde{\Omega}_1^+(S)$ and $\Omega_2(S)$, we can, just as in the proof of (2.3), find k in C_2 orthogonal to $\tilde{y}$ and such that for any $e \in L \cap P^{\perp}$, with $e^2 = -2$, we have $k \notin e^{\perp}$.

Again by [X] exposé X, §4, proposition p. 127, there exists a marked K3 surface (X,φ) of period P. Lemma (2.1), then implies that $\varphi^{-1} \circ S \circ \varphi$ defines a real structure on X.

Conversely, let (X,φ) be a marked K3 surface on which $\varphi^{-1} \circ S \circ \varphi$ defines a real structure.

Let ω be a holomorphic 2-form on X. $F_\infty(\bar{\omega}) = \bar{\omega}^S$ is also a holomorphic 2-form. Let $x = \mathrm{Re}(\omega+\bar{\omega}^S)$ and $y = \mathrm{Im}(\omega+\bar{\omega}^S)$. We can choose $x \in C_1$ and $y \in C_2$. Replacing, if necessary, (x,y) by (-x,-y) we may assume that $x \in C_1^+$.

We have seen that we can always choose a Kähler class k such that $S(k) = -k$ (see II.§1). Hence we can choose $\varphi(k)$ in C_2. Since for any Kähler class we must have $k^\perp \cap \{e \in H^2(X(\mathbb{C}),\mathbb{Z}) \cap H^{1,1}(X) \ / \ e^2 = -2\} = \emptyset$, practically the same argument as the one used in (2.4) shows, that we must have in fact x in $\tilde{\Omega}_1^+(S)$.

It remains to prove that the parametrisation is one to one.

Let (X,φ) and (X',φ') have same period P generated by (x,y), $x \in \tilde{\Omega}_1^+(S)$, $y \in \Omega_2(S)$. Since (X,φ) and (X',φ') have same period they are complex isomorphic by the complex Torelli theorem. More precisely, there exists a unique $w \in W_X$ and a unique isomorphism $f : X' \longrightarrow X$, such that $f^* = \varphi'^{-1}\circ\varphi\circ w$. We will need a lemma.

(2.6) <u>Lemma</u> : <u>Let</u> (X,T) <u>be a real structure on a K3 surface and let</u> W_X <u>be as in</u> (1.10). <u>Then for any</u> $w \in W_X$, $w\circ T^*\circ w^{-1}$ <u>defines a real structure on</u> X <u>if and only if</u> w <u>commutes with</u> T^*, <u>that is</u> $w\circ T^*\circ w^{-1} = T^*$.

<u>Proof</u> : We only need to prove that the condition is necessary. We first note that since T is a real structure, T^* is an isometry and $T^*(H^{1,1}(X)) = H^{1,1}(X)$. In particular $\Delta(X)$ is stable under the action of T^*. Also, since $\langle T^*(x),e\rangle = \langle x,T^*(e)\rangle$, $T^*\circ s_e\circ T^* = s_{T(e)}$ and hence $T^*\circ s_e\circ T^* \in W_X$. In other words, $w \longmapsto T^*w\,T^* = w_T$ is an automorphism of W_X.

Since by hypothesis $w\circ T^*\circ w^{-1}$ defines a real structure, we have by (1.9),

$$w\circ T^*\circ w^{-1}(\mathcal{K}_X) = w\circ w_T^{-1}\circ T^*(\mathcal{K}_X) = -\mathcal{K}_X .$$

We also have $T^*(\mathcal{K}_X) = -\mathcal{K}_X$, hence $w\circ w_T^{-1}(\mathcal{K}_X) = \mathcal{K}_X$.

But W_X operates simply transitively on the chambers ((1.10)), so we must have, $w\circ w_T^{-1} = \mathrm{Id}$ or $w = w_T$. Hence the lemma.

End of proof of (2.5) : Since X and X′ are complex isomorphic, we might as well take X = X′ and f, as above, an automorphism. Assume that both $T^* = \varphi^{-1}\circ S\circ\varphi$ and $T'^* = \varphi'^{-1}\circ S\circ\varphi'$ define real structures.

$f^{*-1}\circ T'^*\circ f^*$ defines also a real structure. But by definition of f, this is just $w^{-1}\circ T^*\circ w$ which is equal to T^* by the lemma. Hence $f\circ T'\circ f^{-1} = T$ and T and T′ are equivalent real structures by I.(1.5).

3. Topology of the real part of a K3 surface and construction of real K3 surfaces.

We first recall that a complex K3 surface is simply connected (see [X] exposé VI) and in particular $H^1(X(\mathbb{C}),\mathbb{Z}/2) = 0$. By Krasnov's result $A_1.7$ this implies, that if X is a real K3 surface with $X(\mathbb{R}) \neq \emptyset$, then X is a GM-surface (see I.(3.11)). Since $h^*(X(\mathbb{C})) = 24$ and $B_2(X(\mathbb{C}) = 22$, we have :

(3.1) $$h^*(X(\mathbb{R}),\mathbb{Z}/2) = 24 - 2\lambda$$

$$\chi(X(\mathbb{R})) = 2b - 20$$

and hence,

(3.2) $$\dim H^0(X(\mathbb{R}),\mathbb{Z}/2) = \frac{2 + b - \lambda}{2}$$

$$\dim H^1(X(\mathbb{R}),\mathbb{Z}/2) = 22 - \lambda - b \quad ,$$

where we have written b for $b_2 = \mathrm{rank}(H^2(X(\mathbb{C}),\mathbb{Z}/2)^G)$ and λ for λ_2, λ_2 defined as in I.(3.4).

Taking into account the restrictions on the possible values of (b,λ), given in chapter II, (2.2), (2.5), (2.6) and (3.8) and the fact that by definition, we must have $\lambda \leqslant \mathrm{Inf}(b,22-b)$, (3.1) yields 64 cases for the possible values of $(h^*(X(\mathbb{R}),\chi(X(\mathbb{R})))$. These are described in the table on next page.

(3.3) Possible values of $(\chi X(\mathbb{R}), h^*(X(\mathbb{R})))$ for a K 3 surface

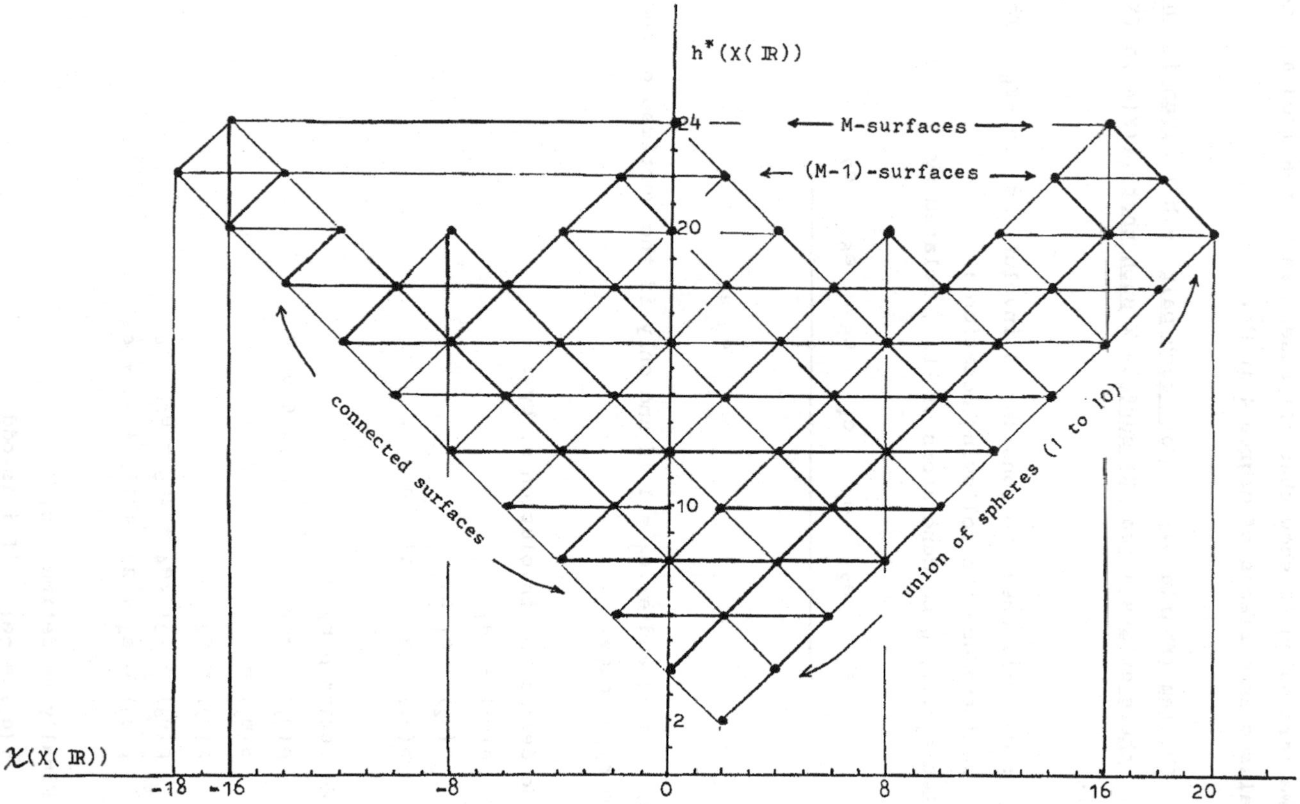

We are going to show that all cases can be realised and can even be realised as surfaces of degree 4 in $\mathbb{P}^3$.

(3.4) <u>Theorem</u> (Kharlamov) : <u>For each pair</u> (χ, h^*) <u>given in the table</u> (3.3), <u>there exists a real K3 surface</u> X <u>such that</u> $(\chi(X(\mathbb{R})), h^*(X(\mathbb{R}))) = (\chi, h^*)$.

<u>Proof</u> : we will need to consider 3 involutions on $-E_8$. To define these, we introduce the following notation :

Let $e_1, \ldots, e_8$ be generators of $-E_8$, related by

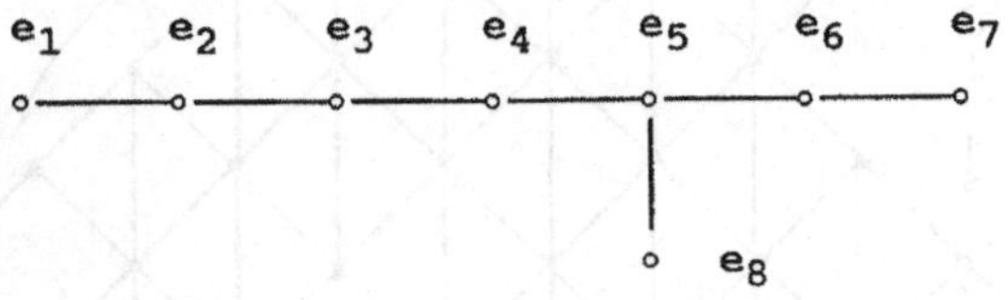

where, $e_i^2 = -2$ and $e_i . e_j = 1$ if and only if the vertices e_i and e_j are joined by an edge.

We define the involution σ, by :

$\sigma(e_1) = -e_1$

$\sigma(e_2) = e_1 + e_2$

$\sigma(e_i) = e_i \quad$ if $i \geqslant 3$.

We define ρ by

$\rho(e_i) = -e_i \quad$ if $i = 4,5,6$ or 8

$\rho(e_1) = e_1$

$\rho(e_2) = e_2$

$\rho(e_3) = e_3 + 2e_4 + 2e_5 + e_6 + e_8$

$\rho(e_7) = e_4 + 2e_5 + 2e_6 + e_7 + e_8$

Finally we define τ by

$\tau(e_i) = -e_i \quad$ if i is odd

$\tau(e_2) = e_1 + e_2 + e_3$

$\tau(e_4) = e_3 + e_4 + e_5$

$\tau(e_6) = e_5 + e_6 + e_7$

$\tau(e_8) = e_5 + e_8$

It is easy to check that these are isometric involutions and that the correspoinding (b,λ) invariants are, (7,1) for σ, (4,2) for e, and (4,4) for τ. We will write σ_i, ρ_i or τ_i for the involution isomorphic to σ, ρ or τ on E_i ($i = 1$ or 2).

Let h_i and h_i' be generators of H_i, $h_i^2 = h_i'^2 = 0$ and $h_i.h_i' = 1$.
Let μ_i be the involution on H_i defined by $\mu_i(h_i) = h_i'$.
We define also $-\mu_i$ to be μ_i composed with -Id. Note that the invariant sub-space of H_i under the action of μ_i (resp. $-\mu_i$) is generated by h_i+h_i' (resp. h_i-h_i') and that $(h_i+h_i')^2 = 2$ (resp. $(h_i-h_i')^2 = -2$). For both, μ_i and $-\mu_i$, the (b,λ) invariant is (1,1).

Let S be an involution on L defined by

$$S = r \oplus s \oplus t \ ,$$

where either, $r = \mathrm{Id}$ or μ_1 on H_1, $s = s_2 \oplus s_3$, with $s_i = -\mathrm{Id}$ or $-\mu_i$ and t is either the involution $\mathfrak{R}$ which exchanges the generators of E_1 and E_2 or $t = t_1 \oplus t_2$ with $t_i = \mathrm{Id}, -\mathrm{Id}, \sigma_i, \rho_i$ or τ_i. Or $r = -\mathrm{Id}$ on H_1, $s = \mathfrak{C}$ the involution that exchanges the generators of H_2 and H_3 and t is as above.

It is easy to check that for each admissible value of (b,λ) on L, or equivalently of (χ, h^*), one can find an S, as above, with exactly these invariants and satisfying to the conditions of theorem (2.3). This proves the theorem.

(3.5) <u>Proposition</u> (Kharlamov) : <u>Let</u> X <u>be a real K3 surface</u>, <u>with</u> $X(\mathbb{R}) \neq \emptyset$. <u>Then</u> $X(\mathbb{R})$ <u>has at most one orientable connected component of Euler characteristic</u> ≤ 0, <u>unless</u> $X(\mathbb{R})$ <u>is the disjoint union of two tori</u> T_1.

For a proof see Risler [X] exposé XIV p. 160-161 (The proof is quite similar to the proof of III.(4.3)).

We will prove in the next section that, for a K3 surface X, $X(\mathbb{R})$ is orientable. This implies, by (3.5), that the invariants (b,λ) completely determine the topology of $X(\mathbb{R})$ except in the cases $(b,\lambda) = (10,10)$, in which case $X(\mathbb{R})$ may be a torus T_1 or empty (see II.(5.2)), and $(b,\lambda) = (10,8)$, in which case $X(\mathbb{R})$ may be the union of two Tori $T_1 \amalg T_1$ or $T_0 \amalg T_2 = S^2 \amalg T_2$.

It is easy to find a smooth quartic in $\mathbb{P}^3$ for which $X(\mathbb{R}) = \emptyset$ and we have given an example (VII.(4.1)) of a quartic with $X(\mathbb{R}) \cong T_1$.

To prove that all cases can occure we only need to prove that there exists real K3 surfaces such that $X(\mathbb{R}) = T_1 \amalg T_1$ and $X(\mathbb{R}) = T_0 \amalg T_2$.

For this we realise the surface as a double covering of $\mathbb{P}^2$ ramified along a sextic (see [X] exposé IV §5).

For the first we choose a real sextic with 3 ovals nested as below

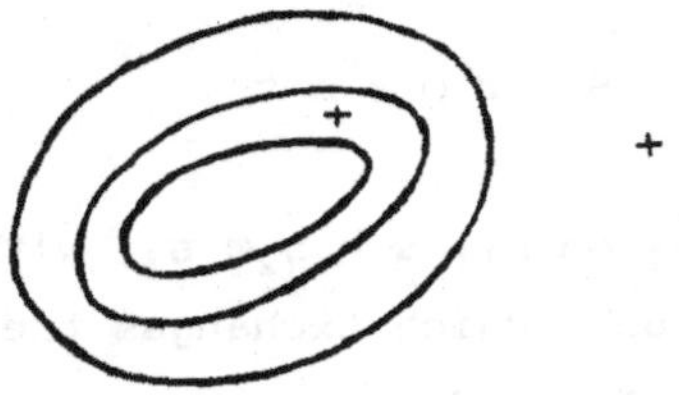

For X we take the double covering of $\mathbb{P}^2$ ramified along this curve and such that the real part is the double covering of the region of $\mathbb{P}^2(\mathbb{R})$ marked + (this is easy to do, see for example [X] exposé XIV p. 157 or the construction we have made in chap. II §4). $X(\mathbb{R})$ is then homeomorphic to $T_1 \cup T_1$.

For the second we apply the same construction to the sextic with 4 ovals arranged in the following way

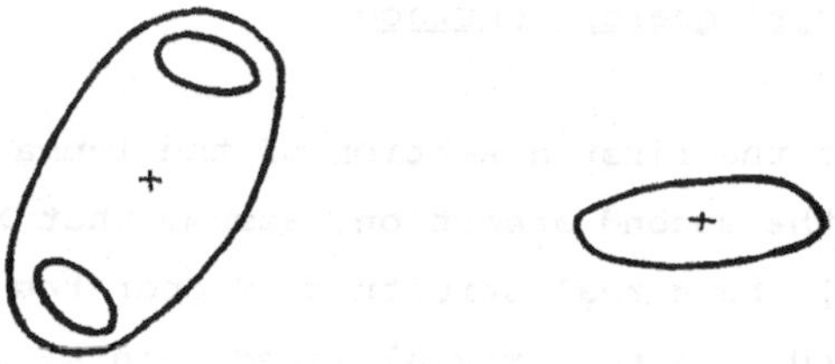

4. **Real quartics in $\mathbb{P}^3$.**

The main result of this section is :

(4.1) Theorem : *Any real K3 surface can be smoothly deformed over $\mathbb{R}$ into a quartic of $\mathbb{P}^3$.*

This theorem has two important consequences.
Since a smooth real quartic surface in $\mathbb{P}^3$ is orientable :

(4.2) Corollary 1 : *For any real K3 surface X, X($\mathbb{R}$) is orientable.*

(4.3) Corollary (Kharlamov). *All the topological types for X($\mathbb{R}$), listed in §3, occure as real parts of real quartic surfaces in $\mathbb{P}^3$.*

For the proof of (4.1) we will need quite a few lemmas and some definitions. We start with a definition.

(4.4) Definition : *A complex K3 surface is special of type g if Pic(X) is cyclic generated by an element α, such that $\alpha^2 = 2g-2$.*

Note that the hypotheses imply that α is non divisible in $H^2(X(\mathbb{C}),\mathbb{Z})$.

(4.5) Lemma : *A K3 surface X special of type 3 is isomorphic to a quartic surface in $\mathbb{P}^3$. Moreover if X is real, with X($\mathbb{R}$) ≠ ∅, then X is*

<u>real isomorphic to a real quartic surface</u>.

<u>Proof</u> : For a proof of the first assertion of the lemma see [X] exposé VI, théorème 2. For the second assertion, assume that X is special of type 3 and let (X,S) be a real structure. S operates on Pic(X) as a group homomorphism and has non trivial fixed points. Since, by hypothesis Pic(X) $\cong \mathbb{Z}$, the action of S on Pic(X) must be the identity. In particular if α is a genertor of Pic(X), $S(\alpha) = \alpha$. By I.(4.5), this implies tht α corresponds to a real invariant divisor. But this divisor is very ample and yields an embedding of X as a quartic surface in $\mathbb{P}^3$. This proves the Lemma.

(4.6) <u>Lemma</u> : <u>Let</u> S <u>be an involution of</u> L <u>satisfying to the hypotheses of</u> (2.3). <u>If</u> $L(1)^G$ <u>contains a hyperbolic plane</u> (<u>that is</u>, a <u>and</u> b <u>such that</u> $a^2 = b^2 = 0$ <u>and</u> $\langle a,b\rangle = 1$) <u>then the set of points in</u> $\Omega(S)$ <u>corresponding to special surfaces of type</u> 3 <u>is dense in</u> $\Omega(S)$.

<u>Proof</u> : Let $P \in \Omega(S)$ be the period of a real K3 surface and let $x \in C_1^+$, $y \in C_2$ (notations an in §2) define P. Let $a \in L(1)^G$ be non divisible and such that $a^2 = 4$. Since $x \in L_{\mathbb{R}}^G$ we have $x \in a^{\perp}$. If $y \in a^{\perp}$ also, then $(x,y) \subset a^{\perp}$, but then, by the proof of théorème 1 of exposé VI of [X], P is the period of special K3 surface of type 3.

With this in mind, we only need to prove that the set of y's in C^2, such that there exists $e \in L(1)^G \cap y^{\perp}$, $e^2 = 4$ and e non divisible, is dense in C^2 To prove this last point we can use the existence of a hyperbolic plane in $L(1)^G$ to follow, word for word, the proof of the corresponding statement over $\mathbb{C}$, given in [X] exposé VI proposition 2.

We are now going to prove that if S is an involution of L, satisfying conditions (2.3), then, unless the (b,λ)-invariants of S are (20,2), $L(1)^G$ contains a hyperbolic plane. By lemma (4.5) and (4.6) this will prove (4.1) except, if the (b,λ) invariants of S are equal to (20,2). We will give a separate argument for this last case.

We will need a notation. Let $(L(1)^G)^{\#} = \{u \in L(1)^G \otimes \mathbb{Q} \ / \ u.L(1)^G \subset \mathbb{Z}\}$. We have $(L(1)^G)^{\#} \cong \mathrm{Hom}_{\mathbb{Z}}(L(1)^G,\mathbb{Z})$. Let $A = (L(1)^G)^{\#}/L(1)^G$. Since the bilinear form on $L(1)^G$ is even, we can define a quadratic form, $q_A : A \longrightarrow \mathbb{Q}/2\mathbb{Z}$, by $q_A(\tilde{u}) = \langle u.u \rangle$ (mod. $2\mathbb{Z}$). We define δ to be 0 if q_A takes its values in $\mathbb{Z}/2\mathbb{Z}$, 1 if not. δ is an invariant of S.

(4.7) Lemma (Nikulin) : If S satisfies to conditions (2.3), then the invariants (b,λ,δ) determine S up to isomorphism.

Proof : By a result of Nikulin (see [Ni] or Dolgachev [Do] Theorem (1.5.1) and (1.5.2)) an involution S on L is determined up to isomorphism by δ, the signature and the determinant of the restriction of the bilinear form to L^G. These last two invariants are uniquely determined by b and λ, hence the Lemma.

(4.8) Corollary : Any involution S of L, satisfying condition (2.5) is isomorphic to one of the involutions introduced in section 3.

Recalling that there are restrictions on δ (see [Do] theorem (1.5.2)) the proof of (4.8) reduces to an elementary computation.

(4.9) Lemma : If the (b,λ) invariant of S, satisfying to condition (2.3) is different from (20,2), then $L(1)^G$ contains a hyperbolic plane.

Proof : By (4.8) we only need to consider $S = r \oplus s \oplus t$, as in §3. If $s = -\mathrm{Id}$ or if $s = s_2 \oplus s_3$ and one of the s_i's equals $-\mathrm{Id}$ then the result is clear. The only other possibility is $r \oplus s = \mu_1 \oplus - \mu_2 \oplus - \mu_3$. In this case $L(1)^G = \langle 2 \rangle \oplus \langle 2 \rangle \oplus (E_1 \oplus E_2)(1)^G$ (where $\langle 2 \rangle$ is the lattice $\mathbb{Z}e$ with $e^2 = 2$). We note that whatever the restriction of S to $E_1 \oplus E_2$ is, δ will be equal to 1. Hence replacing if necessary $\mathfrak{R}$ by $\tau_1 \oplus \tau_2$ (which does not change the invariants (b,λ)), we may assume that there exists e_1 and e_2 in $(E_1 \oplus E_2)(1)^G$ such that $e_1^2 = e_2^2 = -2$ and $e_1.e_2 = 1$. Taking h with $h^2 = 2$, orthogonal to e_1 and e_2 we see that $(h+e_1)$ and $(h-e_2)$ generate a hyperbolic plane.

As noted above this proves theorem (4.1) in all cases, except when the (b,λ) invariants are equal to (20,2). In this last case we note that the restriction of the bilinear form to $L(1)^G$ is positive definite of determinant equal to 4. The only possibility is $L(1)^G = \langle 2\rangle \oplus \langle 2\rangle$. In particular there is a non divisible element of square equal to 4. To conclude the proof we note that $L(1)^G$ does not contain elements of square -2 and that by definition this implies that $\Omega(S)$ is connected.

Bibliographical Notes :

All the results of §3 and §4 are due to Kharlamov [Kh_1], [Kh_2] (see also Risler [X] exposé XIV). The methods used here, and developed in §1 and §2 are slightly different.

It should be noted that Nikulin has also constructed a Moduli space of real K3 surfaces, quite similar to the one given here.

APPENDIX TO CHAPTER VIII

REAL ENRIQUES SURFACES

We will not do here a complete study of real Enriques surfaces, but give a list of possible topological types for the real part, give a method of construction and an example.

We recall that for a complex Enriques surface Y we have, (see [B&P&V] p. 270 and ff), $h^*(Y(\mathbb{C}),\mathbb{Z}/2) = 16$, $B_1 = 0$ (but $B_1(\mathbb{Z}/2) = 1$), $B_2 = 10$ (but $B_2(\mathbb{Z}/2) = 12$) and $h^{0,2} = 0$. By the results of chapter II §2 and §3 this implies the following inequalities

$$(A_8.1) \qquad h^*(X(\mathbb{R})) \leqslant 16$$
$$-8 \leqslant \chi(X(\mathbb{R})) \leqslant 10.$$

Moreover, by II.(2.2) and (2.5) we have :

$(A_8.2)$ $b_2 = 0$ or 8 and $\chi(X(\mathbb{R})) = -8$ or 8 if $h^*(X(\mathbb{R})) = 16$,

and

$(A_8.3)$ $b_2 = 1$, 7 or 9 and $\chi(X(\mathbb{R})) = -6$, 6 or 10 if $h^*(X(\mathbb{R})) = 14$.

From $(A_8.1)$, $(A_8.2)$ and $(A_8.3)$ we obtain the following table, for the à priori possible values of $(\chi(X(\mathbb{R})), h^*(Y(\mathbb{R})))$, when Y is an Enriques surface.

$(A_8.4)$

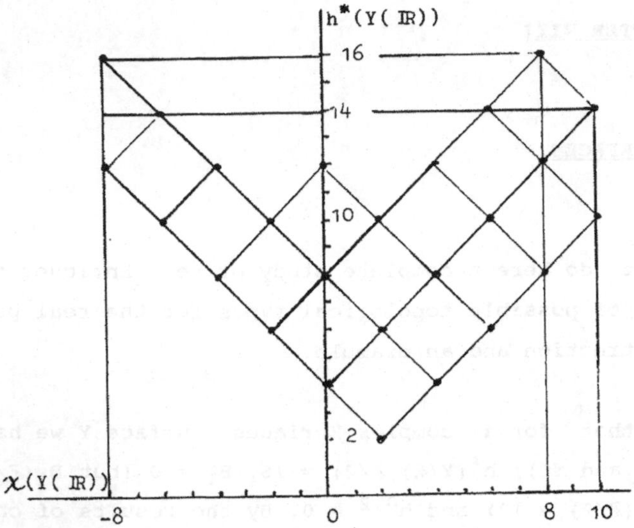

$(A_8.5)$ Remark : It is unclear whether or not, there exists real Enriques surfaces for all these possible cases.

To construct real Enriques surfaces, we will need :

$(A_8.6)$ Theorem : Let Y be real Enriques surface with a real structure (Y,S). Then there exists a real K3 surface X, universal cover of Y, with a real structure (X,S') such that the diagram

$$\begin{array}{ccc} X & \xrightarrow{S'} & X \\ \downarrow & & \downarrow \\ Y & \xrightarrow{S} & Y \end{array}$$

commutes.

Proof : To construct X it is in fact, enough to follow the classical construction used in the complex case. Let $\tilde{K}$ be the total space of the line bundle associated to the canonical sheaf Ω_Y^2 on Y. By composing complex conjugation in the fibres with the action of S on Y we obtain an anti-holomorphic involution $\tilde{S}$ on $\tilde{K}$.

Since Y is an Enriques surface $\tilde{K} \otimes \tilde{K}$ is isomorphic to the trivial bundle. Fix an isomorphism α from $\tilde{K} \otimes \tilde{K}$ to the trivial bundle compatible with $\tilde{S}$ and let

$$X = \{(p.\xi) \in \tilde{K} \ / \ \xi \in \tilde{K}_p,\ \alpha(\xi\otimes\xi) = 1\}.$$

Then X is a K3 surface (see [Gr&Ha] p. 595 or [Be] p. 135), étale covering of Y.

By construction X is stable under the action of $\tilde{S}$. In other words, $\tilde{S}$ induces a real structure S′ on X, which obviously satisfies the conclusions of (1.6).

We use the notations of VIII proof of (3.4).
Let Σ be the involution $-Id \oplus \mathfrak{C} \oplus \mathfrak{R}$ on L, that is, if we write $L = H_1 \oplus H_2 \oplus H_3 \oplus E_1 \oplus E_2$, the involution that is $-Id$ on H_1, exchanges the generators of H_2 and H_3 and the generators of E_1 and E_2.

Let $\Lambda^- = \{x \in L \ / \ \Sigma(x) = -x\}$. Let x and y be linearly independant in $\Lambda^-_{\mathbb{R}} = \Lambda^- \otimes \mathbb{R}$. If for any $e \in \Lambda^-$ with $e^2 = -2$, $e \notin \{x,y\}^{\perp}$ (where $\{x,y\}$ is the plane generated by x and y) there exists a marked K3 surface (X,φ) of period $P = \{x,y\}$, on which Σ induces a holomorphic involution without fixed points. The quotient of X by this involution is an Enriques surface (see [B&P&V] or Horikawa [Ho]).

Let S be an involution on L satisfying to the hypothesis of VIII (2.3) and such that Σ and S commute. If S induces a real structure on the K3 surface X, then passing to the quotient, S induces a real structure on the Enriques surface Y. We apply this to the construction of an example.

(A_8.7) <u>Example of real Enriques surface for which</u> $\chi(X(\mathbb{R})) = -8$ <u>and</u> $h^*(Y(\mathbb{R})) = 16$, <u>that is</u>, $Y(\mathbb{R}) = U_0 \amalg U_{10}$ (<u>see</u> II.(6.7)).

Let $S = Id \oplus -Id \oplus -Id \oplus -Id \oplus -Id$ for the same decomposition into an orthogonal sum as for Σ. Clearly S and Σ commute. Let h_1 and h_1' be generators of H_1 and let $x = h_1 + 2h_1'$. We have $\Sigma(x) = -x$, $S(x) = x$ and x is not orthogonal to an element of L^G of square -2.

Let C_2 be as in VIII (2.5). It is easy to see that $\Lambda_{\mathbb{R}}^- \cap C_2$ is non empty (in fact, it is a space of dimension 10). Let $\Sigma(e) = -e$ and $e^2 = -2$. To prove the existence of the Enriques surface we must show that $e^\perp$ does not contain $\Lambda_{\mathbb{R}}^- \cap C_2$ or that $x \notin e^\perp$. (We can then use the argument, that the e's as above form a discrete family).

Assume first that $S(e) = -e$ and let $y \in \Lambda_{\mathbb{R}}^- \cap C_2$. If $y \in e^\perp$ there exists, since $y^2 > 0$, $a \in \mathbb{R}$ such that $(ay+e)^2 > 0$. Since y and e are both in $\Lambda_{\mathbb{R}}^- \cap L_{\mathbb{R}}(1)^G$, $(ay+e) \in \Lambda_{\mathbb{R}}^- \cap L_{\mathbb{R}}(1)^G$ and hence $(ay+e) \in \Lambda_{\mathbb{R}}^- \cap C_2$ and is not orthogonal to e.

If $S(e) \neq -e$, there exists, since $L = L^G \oplus L(1)^G$ (this because of the hypothesis on S), $e' \in L^G$ and $e'' \in L(1)^G$ such that $e = e'+e''$. We have $e'.y = 0$ and hence $e''.y = 0$. If e" is isotropic we have $(e')^2 = -2$ and hence $x \notin e'^\perp$ and $x \notin e^\perp$. If $(e'')^2 \neq 0$, we can apply to e" the argument applied above to e, in the case $S(e) = -e$.

To conclude we note that by the choice of S we have constructed a K3 surface with $X(\mathbb{R}) = S^2 \amalg T_{10}$. Hence the assertion on $Y(\mathbb{R})$.

In general we do not have $L = L^G \oplus L(1)^G$, and this method turns out to be of delicat use.

GLOSSARY OF NOTATIONS

G	$= \mathrm{Gal}(\mathbb{C}\|\mathbb{R})$	V
S	generator of G..............................	V
j_n, j	complex conjugation..........................	1
f^σ	conjugate function..........................	1
$X^\sigma(\mathbb{C})$	complex conjugate variety....................	1,2
$\bar{\mathcal{O}}_{X(\mathbb{C})}$	sheaf of anti holomorphic functions.........	2
X^σ	conjugate scheme.............................	2
(X,S)	real structure...............................	3,8
f^S	....	3
X_o	real model...................................	4,8
A^G	fixed part under a G-action.................	5,14
$X_o(\mathbb{C})^G$	real part....................................	5
$X_o(\mathbb{R})$	real part....................................	5
$X(\mathbb{R})$, $(X,S)(\mathbb{R})$	real part....................................	5,8
$\mathcal{L}^S$	image of a sheaf $\mathcal{L}$ under S..................	6
$P_a(Y)$	arithmetic genus.............................	7
$q(Y)$	irregularity.................................	7
$Y^S = S(Y)$	image by S of a sub-variety..................	7
$\chi(X(\mathbb{R}))$	topological Euler characteristic............	9
B_i	Betti number of $X(\mathbb{C})$.........................	9
b_i	$\dim H^i(X(\mathbb{C}),\mathbb{Q})^G$..............................	9,12
$\chi(X(\mathbb{C}))$	topological Euler characteristic............	10
F_∞	$= S^* \otimes \mathrm{Id}$..............................	10
$H^{p,q}(X(\mathbb{C}))$	Hodge spaces.................................	10
$h^{p,q}$	Hodge numbers................................	12
a_m	Smith number.................................	13
λ_m	Comessatti characteristic....................	14
(b,λ)	(b,λ)-invariants.............................	15
$H^i(G,A)$	Galois cohomology groups.....................	15
$S(D), D^S$	image by S of a divisor......................	19
$\mathcal{O}(D)$	sheaf of a divisor...........................	19
$D \sim D'$	linear equivalence...........................	19
$D \approx D'$	algebraic equivalence........................	20
$\mathrm{Pic}(X)^G$	real part of the Picard group................	20

INDEX

BIBLIOGRAPHY

[Al] A.A. ALBERT : Symmetric and alternate matrices in an arbitrary field I ; Trans. Amer. Math. Soc. 43, 386-436 (1938).

[B&P&V] W. BARTH, C. PETERS, A. VAN DE VEN : Compact Complex Surfaces ; Ergebnisse der Mathematik, Springer, Berlin, Heidelberg 1984.

[Be] A. BEAUVILLE : Surfaces Algébriques Complexes ; Astérisque 54, Paris 1978.

[Be & To] R. BENEDETTI, A. TOGNOLI : Remarks and counterexamples in the theory of real algebraic vector bundles and cycles ; in Géométrie Algébrique Réelle et Formes Quadratiques , 198-211; Lecture Notes in Math. 959, Springer, Berlin-Heidelberg 1982.

[Bo & Mu] E. BOMBIERI, D. MUMFORD : Enriques classification in char. P, II ; in Complex Analysis and Algebraic Geometry , 23-42; Iwanami-Shoten, Tokyo, 1977.

[Bo & Ha] A. BOREL, A. HAEFLIGER : La classe d'homologie fondamentale d'un espace analytique ;Bul. Soc. Math. France 83, 461-513 (1961).

[B&C&R] J. BOCHNAK, M. COSTE, M.F. ROY : Géométrie Algébrique Réelle ; Ergebnisse der Mathematik, Springer, Berlin, Heidelberg 1987.

[B&K&S] J. BOCHNAK, W. KUCHARZ, M. SHIOTA : The divisor class group of global real analytic, Nash or rational regular functions; in Géométrie Algébrique Réelle et Formes Quadratiques, 218-248;Lecture Notes in Math. 959, Springer, Berlin, Heidelberg, 1982.

[Bou] N. BOURBAKI ; Groupes et Algèbres de Lie chap. 4,5,6 ; Hermann, Paris 1968.

[Br] L. BROCKER : Reelle Divisoren ; Arch. Math. 35, 140-143 (1980).

[Ca] H. CARTAN : Quotient d'un espace analytique par un groupe d'automorphismes ; in Algebraic Geometry and Topology, a Symposium in Honor of S. Lefschetz , 90-102; Princeton Univ. Press, Princeton 1957.

[Col] J.-L. COLLIOT-THÉLÈNE : Arithmétique des variétés rationnelles et problèmes birationnels ; Proc. Inter. Congress of Maths. Berkeley, 1986.

[Co&Sa] J.-L. COLLIOT-THÉLÈNE, J.-J. SANSUC : Variétés de première descente attachées aux variétés rationnelles ; C.R. Acad. Sc. 284 Série I, 967-970, (1977).

[Co_1] A. COMESSATTI : Fondamenti per la geometria sopra le superficie rationali del punto di vista reale ; Math. Ann. 73, 1-72 (1912).

[Co_2] A. COMESSATTI : Sulle varietà abeliane reale I e II ; Ann. Mat. Pura. Appl. 2, 67-106 (1924) and 4, 27-71 (1926).

[Co_3] A. COMESSATTI : Sulla connessione delle superficie algebraiche reali ; Ann. Mat. Pura Appl. 5, 299-317 (1927-1928).

[De_1] P. DELIGNE : Valeurs des fonctions L et périodes d'intégrales ; proc. Symp. pure Math. 33, 313-346 ; Amer. Math. Soc. Providence, 1979.

[De_2] P. DELIGNE : Le théorème de Noether ; in SGA 7 II, 328-340 ; Lecture Notes in Math. 340, Springer Berlin-Heidelberg 1973.

[De&Mu] P. DELIGNE, D. MUMFORD : The irreducibility of the space of curves of given genus ; Pub. Math. I.H.E.S. 36, 75-109 (1969).

[Dem] M. DEMAZURE : Surfaces de Del Pezzo ; in Séminaire sur les Singularités des Surfaces , 21-69; Lecture Notes in Math. 277, Springer, Berlin-Heidelberg, 1980.

[D_i] J. DIEUDONNE : Cours de Géométrie Algébrique ; P.U.F., Paris, 1974.

[Do] J. DOLGACHEV : Integral quadratic forms : Applications to Algebraic Geometry ; Seminaire Bourbaki n° 611 (1982/83).

[Fl_1] E. FLOYD : On periodic maps and Euler characteristic of associated spaces ; Trans. Amer. Math. Soc. 72, 138-147, (1952).

[Fl_2] E. FLOYD : Periodic maps via Smith theory ; in Seminar on Transformation Groups, 35-48 ; Princeton Univ. Press, Princeton 1960.

[Gr&Ha] P GRIFFITHS, J. HARRIS : Principles of Algebraic Geometry ; J. Wiley & Sons ; New York 1978.

[Gr_1] A. GROTHENDIECK, Sur quelques points d'algèbre homologique; Tôhoku Math. J. 9 ; 119-221 (1957).

[Gr_2] A. GROTHENDIECK : Techniques de descente et théorèmes d'existences en géométrie algébrique I et V ; Séminaire Bourbaki Exp. 190 (1959/60) et Exp. 232 (1961/62).

[Gu] D.A. GUDKOV : The topology of real projective algebraic manifolds ; Russian Math. Surveys 29, 1-79 (1974).

[Gu&Kr] D.A. GUDKOV, A.D. KRAKNOV : On the periodicity of the Euler characteristic of real algebraic (M-1)-manifolds ; Funct. Anal. and Appl. 7, 15-19 (1973).

[Ha] R. HARTSHORNE : Algebraic Geometry ; Graduate Texts in Math. 49, Springer, Berlin Heidelberg, 1977.

[Hi] M. HIRSCH : Differentiel Topology ; Graduate Text in Math. 33, Springer New-York 1976.

[Hz] F. HIRZEBRUCH : The signature of ramified coverings ; in Global

Analysis (in honour of Kodaira) ; Univ. Press Tokyo, 1969.

[Ho] E. HORIKAWA : On the periods of Enriques surfaces I and II, Math. Ann. 234, 73-88 (1978) and 235, 217-246 (1978).

[Is_1] V.A. ISKOVSKIH : On birational forms of rational surfaces ; Amer. Math. Soc. Transl. 84, 119-136 (1967).

[Is_2] V.A. ISKOVSKIH : Rational surfaces with a pencil of rational curves ; Math. USSR. Sb. 3, 563-587 (1967).

[Kh_1] V.M. KHARLAMOV : Additional congruence for the Euler characteristic of real algebraic manifolds of even dimensions ; Funct. Anal. Appl. 9, 134-141 (1975).

[Kh_2] V.M. KHARLAMOV : The topological type of non-singular surfaces in RP^3 of degree four ; Funct. Anal. Appl. 10, 295-305 (1976).

[Ko] K. KODAIRA : On compact analytic surfaces II ; Annals of Math. 77, 563-626 (1963).

[Kn] M. KNEBUSCH : On algebraic curves over real closed field I and II ; Math. Z. 150, 49-70 (1976) and 151, 189-205 (1976).

[Kr] V.A. KRASNOV : Harnack-Thom inequalities for mappings of real algebraic varieties ; Math. USSR izvestiya 22, 247-275 (1984).

[Ma_1] Y. MANIN : Rational surfaces over perfect fields ; Amer. Math. Soc. Transl. 84, 137-186 (1969) and Math. USSR-SB.,1, 141-168, (1967).

[Ma_2] Y. MANIN : Cubic Forms ; North Holland, Amsterdam 1974.

[Mas] W. MASSEY : Homology and Cohomology Theory ; Marcel Dekker New York 1978.

[Mi&Hu] J. MILNOR, D. HUSEMOLLER : Symmetric Bilinear Forms; Springer,

Berlin-Heidelberg 1973.

[Mi&St] J. MILNOR, J. STASHEFF : Characteristic Classes ; Ann. of Math. Studies, Princeton Univ. Press, Princeton 1974.

[Mu_1] D. MUMFORD : Lectures on Curves on an Algebraic Surface ; Annals of Math. Studies 59, Princeton Univ. Press, Princeton 1966.

[Mu_2] D. MUMFORD : Abelian Varieties ; Oxford Univ. Press, Oxford 1974.

[Ne] A. NERON : Modèles minimaux des variétés abéliennes sur les corps locaux et globaux ; Publi. Math. I.H.E.S., 21, 361-483 (1964).

[Ni] V.V. NIKULIN : Involutions of integral quadratic forms and their applications to real algebraic geometry ; Math. USSR Izvestiya 22, 99-172 (1984).

[Pj&Sh] I.I. PJATSCKII-SHAPIRO, I.R. SHAFAREVICH : A Torelli theorem for algebraic surfaces of type K3 ; Math. USSR Izvestiya 3, 547-588 (1971).

[Ra] M. RAYNAUD : Spécialisation du foncteur de Picard ; Publ. Math. I.H.E.S. 38, 27-76 (1970).

[Ri_1] J.-J. RISLER : Sur le 16ème problème de Hilbert : un résumé et quelques questions ; Pub. Math. de l'Univ. Paris VII n° 9 (1980).

[Ri_2] J.-J. RISLER : Type topologique des surfaces algébriques réelles de degré 4 dans RP^3 ; in Séminaire sur la Géométrie des surfaces K3 : Modules et Périodes ; Astérisque 126 Paris 1985.

[Rh] K. ROHN : Die Maximalzahl und Anordnung des Ovale bei der ebenen Kurve 6 Ordnung und bei der Fläche 4 Ordnung ; Math. Ann. 73, 177-229 (1913).

[Ro] V.A. ROKHLIN : Congruence modulo 16 in Hilbert's 16th problem ; Funct. Anal. Appl. 6, 301-306 (1972).

[Seg] B. SEGRE : The Non-Singular Cubic Surfaces ; Clarendon Press, Oxford 1942.

[Se&Kn] J.G. SEMPLE, G.T. KNEEBONE : Algebraic Projective Geometry ; Oxford Univ. Press, Oxford 1952.

[Se_1] J.-.P. SERRE : Géométrie algébrique et géométrie analytique ; Ann. Inst. Fourier 6, 1-42 (1956).

[Se_2] J.-P. SERRE : Corps Locaux ; Hermann, Paris 1962.

[Se_3] J.-P. SERRE : Cours d'Arithmétique ; P.U.F., paris 1970.

[Sha] I.R. SHAFAREVICH : Algebraic Surfaces ; Proceedings of the Steklov Institute of Math. ; Amer. Math. Soc. Providence 1967.

[Sh_1] G. SHIMURA : On the field of rationality for an abelian variety ; Nagoya Math. J. 45, 167-178 (1972).

[Sh_2] G. SHIMURA : On the real points of an arithmetic quotient of a bounded symmetric domain ; Math. Ann. 215, 135-164 (1975).

[Si_1] R. SILHOL : Real abelian varieties and the theory of Comessatti ; Math. Z. 181, 345-364 (1982).

[Si_2] R. SILHOL : A bound on the order of $H_{n-1}^{(a)}(X,\mathbb{Z}/2)$ on a real algebraic variety ; in Géométrie Algébrique Réelle et Formes Quadratiques ; Lecture Notes in Math. 959, 443-450, Springer, Berlin-Heidelberg 1982.

[Si_3] R. SILHOL : Real algebraic surfaces with rational or elliptic fiberings ; Math. Z. 186, 465-499 (1984).

[Si_4] R. SILHOL : Bounds for the number of connected components and the first Betti number mod 2 of a real algebraic surface; Compositio Math. 60, 53-63 (1986).

[Si_5] R. SILHOL : Cohomologie de Galois des variétés algébriques réelles : application aux surfaces rationnelles ; Bul. Soc. Math. France 115, 107-125 (1987).

[Ta] J. TATE : Algebraic cycles and poles of Zeta functions ; in Arithmetical Algebraic Geomery, Perdue Univ. Conference (1963) ; Harper & Row, New York 1963.

[Vi_1] O.Y. VIRO : Curves of degree 7, curves of degree 8 and the Ragsdale conjecture ; Soviet Math. Dokl. 22, 566-570 (1980).

[Vi_2] O.Y. VIRO : Construction of real algebraic surfaces with many components ; Soviet Math. Dokl. 20, 991-995 (1979).

[We_1] A. WEIL : The field of definition of a variety ; Amer. J. Math. 78, 509-524 (1956).

[We_2] A. WEIL : Reduction des formes quadratiques ; Séminaire H. Cartan (1957/58) Exp. 1.

[We_3] A. WEIL : Remarques sur un mémoire d'Hermite ; Arch. Math., 197-202 (1954).

[Wi] G. WILSON : Hilbert's sixteenth problem ; Topology 17, 53-73 (1978).

[Wit] E. WITT : Zerlegung reeller algebraisher Functionen in Quadrate. Schiefkörper über reellem Funktionenkörper ; J. Reine Angew. Math. 171, 4-11 (1934).

[X] Séminaire Palaiseau : Géométrie des Surfaces K3 : Modules et Périodes ; Astérisque 126, Soc. Math. de France, Paris 1985.